CDM 2007

Q&A

Pat Perry

ELSEVIER

AMSTERDAM • BOSTON • HEIDELBERG • LONDON
NEW YORK • OXFORD • PARIS • SAN DIEGO
SAN FRANCISCO • SINGAPORE • SYDNEY • TOKYO

Butterworth-Heinemann is an imprint of Elsevier

Butterworth-Heinemann is an imprint of Elsevier
Linacre House, Jordan Hill, Oxford OX2 8DP, UK
30 Corporate Drive, Suite 400, Burlington, MA 01803, USA

Notice
No responsibility is assumed by the publisher for any injury and/or damage to persons
or property as a matter of products liability, negligence or otherwise, or from any use
or operation of any methods, products, instructions or ideas contained in the material
herein. Because of rapid advances in the medical sciences, in particular, independent
verification of diagnoses and drug dosages should be made

British Library Cataloguing-in-Publication Data
A catalogue record for this book is available from the British Library.

Library of Congress Cataloging-in-Publication Data
A catalog record for this book is available from the Library of Congress

ISBN: 978-0-7506-8708-9

For information on all Butterworth–Heinemann publications
visit our web site at books.elsevier.com

Typeset by Charon Tec Ltd (A Macmillan Company), Chennai, India
www.Charontec.com

Printed in Great Britain

08 09 10 11 12   10 9 8 7 6 5 4 3 2 1

Contents

18 Asbestos

1

The Construction (Design and Management) Regulations 2007 – The Regulations Explained

1.1 What are the CDM Regulations and what is their purpose?

The CDM Regulations are the common name for the Construction (Design and Management) Regulations 2007, a set of Regulations which came into force on 6 April 2007.

The Regulations replace the Construction (Design and Management) Regulations 1994 (and amendments) and the Construction (Health, Safety & Welfare) Regulations 1996.

The purpose of the Regulations is to improve the health and safety record on construction sites by requiring all parties involved in a construction project to take responsibility for health and safety standards. By requiring better planning, design and management of a construction project, it is believed that unacceptably high accident and fatality rates will be reduced.

The Regulations are also intended to help reduce the high incident rate of occupational ill-health which is common in the construction industry e.g. chronic respiratory conditions, muscular–skeletal conditions, industrial deafness, industrial dermatitis and so on.

Those who could create health and safety risks have been made responsible for considering and controlling them during all stages of the project – conception, design, planning, construction work, future maintenance and use of the building, including demolition.

1.2 Why have the 1994 Regulations been replaced?

The CDM Regulations 1994 were instrumental in improving the overall health and safety on construction sites but they were deemed relatively ineffective in reducing design risks, changing attitudes to the importance of planning and managing projects and so did not bring the improvements in health and safety expected.

In particular, clients and designers were slow to accept their duties and failed to understand the intent of the Regulations regarding hazard elimination, etc.

The Planning Supervisor role was seen by most parties involved in a construction project as unnecessary and ineffective, and there was very little "planning" or "supervision" of projects by them.

The 1994 Regulations also generated huge amounts of paperwork and it seemed that the volume of paperwork generated for a project would indicate how effectively they were complying with CDM.

Often, all duty holders under the 1994 Regulations would hide behind the paperwork and little improvement was seen in relation to managing, communicating and co-ordinating risk on a construction project.

The old Regulations were seen as complex and bureaucratic and divorced from the Regulations governing site safety i.e. the Construction (Health, Safety and Welfare) Regulations 1996.

There were also concerns raised by the EU Commission that the UK had not adequately adopted the relevant EU Directive and improvements were being required to ensure consistency across all EU Countries.

1.3 Do the CDM Regulations apply to all construction projects?

Yes.

The 2007 Regulations apply to all construction works irrespective of how many operatives are engaged on the project. The 1994 Regulations exempted construction projects where there were five or fewer operatives on site. The 2007 Regulations have removed this exemption because health and safety is as much a concern on smaller sites as it is on larger sites.

Not all of the 2007 Regulations apply to all construction projects – Part 3 which covers additional duties where a project is notifiable applies only if the construction phase lasts more than 30 days or more than 500 person days.

Most duties remain on clients, designers and contractors regardless of notification.

1.4 What is "construction work" under the CDM Regulations?

Construction work means the carrying out of building, civil engineering or engineering construction work.

The definition includes:

- The construction, alteration, conversion, fitting out, commissioning, renovation, repair, upkeep, redecoration or other maintenance (including cleaning which involves the use of water or an abrasive at high pressure or the use of substances classified as corrosive or toxic) de-commissioning, demolition or dismantling of a structure.
- The preparation for an intended structure, including site clearance, exploration, investigation (but not site survey), excavation, laying and installing the foundations of the structure.
- The assembly on site of pre-fabricated elements to form a structure or the disassembly on site of pre-fabricated elements which immediately before such disassembly formed a structure.
- The removal of a structure or part of a structure or any product or waste resulting from demolition or dismantling of a structure or from the disassembly of pre-fabricated elements which, immediately before disassembly, formed a structure.
- The installation, commissioning, maintenance repair or removal of mechanical, electrical, gas, compressed air, hydraulic, telecommunications, computer or similar services which are normally fixed within or to a structure.

1.5 What is not "construction work"?

The following activities are generally not classed as construction work:

- General horticultural work and tree planting
- Archaeological investigations
- Erecting and dismantling of marquees
- Erection and dismantling of lightweight partitions to divide open plan offices
- Creation of exhibition stands and displays
- Erection of scaffolds for support or access for work activities which are not classed as construction works
- Site survey works e.g. taking levels, assessing soil types, examining structures
- Work to or on ships
- On-shore fabrication of elements for off-shore installations
- Factory manufacture of items for use on construction sites.

1.6 The CDM Regulations refer to key appointments which must be made on a construction project. What does this mean?

The CDM Regulations identify key duty holders who have responsibilities for ensuring that health and safety matters are addressed during construction projects. They are:

(1)   The Client
(2)   The Designer

(3)   The CDM Co-ordinator (only for notifiable projects)
(4)   The Principal Contractor (only for notifiable projects)
(5)   Contractors.

The Client is anyone for whom a construction project is carried out.

The Designer is anyone who carries on a trade, business or undertaking in connection with which he:

- Prepares a design, or
- Arranges for any person under his control to prepare a design,

relating to a structure or part of a structure.

The CDM Co-ordinator is a competent person or body with the overall responsibility for co-ordinating health and safety aspects of the design and planning stage and advising the client on all health and safety matters.

The Principal Contractor is appointed on notifiable projects and must be a Contractor. They must take responsibility for all site-specific safety issues, including ensuring that Contractors and Sub-Contractors are competent and have resources to carry out the work safely, that a Health and Safety Plan is developed. Principal Contractors are also responsible for providing information, training and consultation with employees, including the self-employed.

Contractors are those who manage or carry out construction work.

1.7 The CDM Regulations refer to a "notifiable project". What is this?

The Regulations require certain projects to be notified to the Health and Safety Executive (HSE).

A construction project is notifiable to the HSE area office when:

- It will, or is expected to, last more than 30 days, or
- It will, or is expected to, involve more than 500-person days.

The HSE requires certain information which is outlined in Schedule 1 of the CDM Regulations. As long as the relevant information is given it can be supplied in any format but in order to facilitate notification the HSE have produced Form F10 (rev.) which can be used for all projects.

An example of Form F10 (rev.) is shown in Appendix 5A.

1.8 When does the construction phase start and are weekends and bank holidays counted?

The construction phase for the purposes of notification to the HSE is from the day "construction works" start.

Remember, *site clearance* constitutes construction works and must be included in the calculation.

If construction work is programmed to take place on Saturdays and Sundays and on any Bank Holiday, no matter how long the working shift will be, these days must be counted as "construction days".

1.9 What is a "person day"?

A "person day" is any day or part of a day (no matter how short) when some-one is expected to carry out construction work. A person day relates to *one* individual and includes Site Agents, Foreman and Supervisors.

They do not actually have to be carrying out any physical work to be involved in "construction work" – if they are managing the project they are included as a "person day".

1.10 Why do the HSE need to know about these projects?

The HSE enforces health and safety laws in the construction industry. Projects lasting over 30 days are considered to be substantial building or refurbishment projects where the risks to health and safety of operatives and others can be high. The HSE has always been made aware of construction projects so that they can plan their inspection programme of enforcing the laws.

The HSE receive over 500,000 project notifications per year and use the notification process to help them target their on-site inspection priorities. In addition, they will review the F10's to try to influence the design process, enquiring about the provision for health and safety where appropriate.

1.11 What does the HSE do when they receive all these F10's?

The Construction division of the local HSE office records all forms received and allocates the projects to individual inspectors. The HSE inspectors base their routine inspections on the type of projects notified and will prioritize projects into perceived risk categories e.g. those where falls from height are likely, or which involve excavations and so on.

The Construction Inspector, or the Administration Officer, will look at key dates on the form. They are interested to note when the form is signed by the Client (or on his behalf) and received by them, and when works are proposed to start on site.

The time allowed for preparation and planning on the project will be of critical importance to the HSE.

If there are only a few days between the HSE receiving the form and works starting on site they will investigate because the essence of CDM is to involve the CDM Co-ordinator (and Designers) in the planning process of a project in respect of health and safety and to allow the Principal Contractor adequate time to plan and prepare for the start on site.

They will want to establish whether:

- The Client appointed a CDM Co-ordinator early enough in the project i.e. as soon as he had information about the project and the construction work involved
- The CDM Co-ordinator notified the project to the HSE as soon as was practicable after their appointment.

A Client who fails to appoint a CDM Co-ordinator early enough in the project is guilty of an offence. Likewise, a CDM Co-ordinator who fails or delays in notifying a project unreasonably, is guilty of an offence.

It is not a valid excuse for either party to rely on the fact that financial resources had not been released for the project and the project was therefore not "live".

Case Study

A Developer was undertaking "core and shell" works for new retail units and had notified HSE in the usual way. The shell unit went under offer to a restaurant company who commissioned their Designers to fit out the shell. The Developers work over-ran the timescales and legal complications delayed the Client purchase. The Client appointed the CDM Co-ordinator as soon as the legal work had been completed and the shell building was handed over. The Client wanted works of fitting out to start as soon as possible. The CDM Co-ordinator notified the HSE on Form F10, indicating that works were due to start on site in approximately 2 weeks.

The HSE wrote to the CDM Co-ordinator to enquire at what date they were appointed and why notification had only just taken place. They asked to see letters of appointment for the CDM Co-ordinator with a view to determining whether the Client had failed in their duty to appoint the CDM Co-ordinator "as soon as practicable after information about the project becomes available". Also the Inspector wanted to know what time had been allowed for preparation and planning by the Principal Contractor – 2 weeks seemed inadequate to him for the size of the project.

If the Client had complied with his legal duty of an early appointment, the HSE wanted to establish whether the CDM Co-ordinator was dilatory in making the notification, thereby breaching their duties under Regulation 21.

In this instance, the HSE advised that in their view, the Client had appointed the CDM Co-ordinator too late in the design process and that in future, the Client should appoint the CDM Co-ordinator as soon as an intended purchase had been agreed.

1.12 What happens if a project was originally going to last less than 30 days (or 500-person days) and not be notifiable but due to unforeseen circumstances it will now take longer?

First, you must review *why* you thought it would be lasting less than 30 days in the first place. The Regulations state that where it is reasonable to assume that a project may take longer than 30 days (or 500-person days) it must be notified. The HSE will view seriously any intent to evade the Regulations by avoiding notification.

If the project is only to over-run by a few days the HSE will not expect retrospective notification.

If the project will last a few more weeks, the CDM Co-ordinator should notify the HSE office on Form F10 and include a covering letter explaining the reasons for the project over-run and the late notification.

1.13 Is demolition work covered by CDM 2007?

Yes.

Demolition works are construction works and all of the Regulations apply except Part 3 – notifiable projects – when the works last for less than 30 days or involve less than 500-person days.

If demolition works are expected to last for more than 30 days or involve more than 500-person days then the project must be notified to the HSE and the Client must appoint the appropriate duty holders i.e. CDM Co-ordinator and Principal Contractor.

Specifically under CDM 2007, demolition works are required to have a written plan showing how danger is to be avoided. This plan is to be drafted by those in control of the works (Regulation 29).

1.14 What does demolition and dismantling work include?

Demolition is taken to mean the deliberate pulling down, destruction or taking apart of a structure, or a substantial part of the structure.

Dismantling is the taking down, or taking apart, of all, or a substantial part of a structure, and includes situations where the structure is carefully taken down for re-use.

A "structure" is defined as:

"Any building, steel or re-enforced concrete structure (not being a building), railway line or siding, tramway line, dock, harbour, inland

navigation, tunnel, shaft, bridge, viaduct, waterworks, reservoir, pipe or pipeline (whatever in either case it contains or is intended to contain) cable, aqueduct, sewer, sewerage works, gasholder, road, airfield, sea defence works, river works, drainage works, earthworks, lagoon, dam, wall, caisson, mast, tower, pylon, underground tank, earth retaining structure, or structures designed to preserve or alter any natural feature, and any other structure similar to the forgoing", or

"Any formwork, false work, scaffold or other structure designed or used to provide support or means of access during construction work".

The CDM Regulations apply to demolition and dismantling works irrespective of whether the project is notifiable and regardless of how many operatives are involved in undertaking the work.

1.15 Is there a difference between "demolition" and "dismantling" works?

Demolition is the deliberate pulling down, destruction or taking apart of all, or a substantial part of the structure.

Dismantling is the taking down, or taking apart, of all, or a substantial part of a structure.

Dismantling for re-erection or re-use will be demolition for the purposes of CDM.

The formation of openings for windows, doors and services are not in themselves demolition works.

The removal of cladding, roof tiles or scaffolding is not, in itself, demolition or dismantling works unless included in or combined with other building operations.

Case Study

A small retailer is having a brick outhouse demolished to make way for additional car parking. The building is single storey, single skin brickwork under a flat roof. The local builder can undertake the work in 2 days with 3 men on site.

The works involved fall within the definition of construction work and irrespective of their duration and number of operatives on site the works come under the control of CDM.

The Client – the Shop Owner and the Contractor have to comply with all the requirements of CDM except Part 3: Notifiable Projects.

He appoints a competent contractor and discusses with him the use of the rear yard, the previous use of the outhouse and passes to the contractor any information as to how it was constructed.

The Contractor makes notes that the rear yard is used for access to the shop by delivery vehicles but it is not used as staff access, nor does anyone else have rights of access. The information on the building to be demolished is brick, single skin, and flat-felt roof. No services are connected to the out building, no asbestos was present and the structure to be demolished is straightforward. To comply with Regulation 29 a short written Plan of Works or Method Statement is produced by the Contractor and agreed with by the Client.

He stipulates that the rear yard will be cordoned off, that the shop owner will re-schedule deliveries for later in the week, after demolition, that the skip will be positioned away from the rear emergency exit door, that his operatives will be able to use the shop WC and tea-making facilities, that waste will be removed by a licensed disposal company. In addition, the builder prepares a brief Method Statement covering the sequence of demolition, identifying hazards and risks e.g. dust inhalation, falling from height, manual handling and so on.

The shop owner is given a copy of the Health and Safety Plan and it is agreed before works start.

1.16 How do the CDM Regulations apply to Term Maintenance contracts?

If the works are "construction works" then CDM Regulations apply to those aspects of the project. Therefore, any painting, renewals, repairs, redecoration, maintenance, improvements, etc. are likely to be construction works – it is a wide ranging definition.

If the works are "construction works" and last more than 30 days or involve more than 500-person days then they will be notifiable and all CDM Regulations will apply.

Where some of the works are outside of the Regulations and some within it, it would be advisable to apply CDM to all of the works, although any management arrangements could clearly outline how both aspects of the project would be managed.

Projects involving term maintenance must be reviewed individually.

1.17 Does CDM apply to emergency works?

In any emergency, the first priority is to make sure that the premises or structure is safe and without imminent risk to the health and safety of members of the public and others.

Once the making safe has been carried out then CDM Regulations should be applied i.e. the Client should make the appointment of a CDM Co-ordinator and Principal Contractor as soon as is practicable.

Designers, once appointed, must fulfil their responsibilities under Regulation 11.

The Construction Phase Health and Safety Plan should be developed as soon as is practicable and if time does not allow a written Plan to be completed before the emergency works take place, verbal discussions and agreement should be reached regarding key health and safety issues. If possible, the key issues should be written down and the body of the Plan can then be developed as the works progress.

Emergency works that are likely to be substantial will inevitably fall under the CDM Regulations. It would be wise to consult the local HSE Office about any plans, etc. even if there is no time to prepare the paperwork.

1.18 Who are domestic Clients?

Domestic Clients are Clients who have work done which does not relate to any trade, business or other undertaking. This is usually someone who commissions work on their own home, or the home of a family member.

No duties are placed on domestic Clients by CDM. Those working for the domestic Client may have duties.

1.19 Do the Regulations apply to domestic house building or repairs?

If you commission a builder to build you a house for your own occupation then CDM Regulations do not apply, with the exception that Designers have to comply with their duties to design safely and Contractors have to comply with Part 4 of the Regulations. A domestic client has no "client" responsibilities under the Regulations.

If you have a new conservatory, extension, loft conversion or similar then CDM does not apply. If you have major renovations carried out, then CDM does not apply. Designers will have to comply with Regulation 11 and any contractors will have to ensure that health and safety requirements are met.

If you commission a house to be built to sell it to someone else, then you are likely to be a developer and CDM Regulations will apply, including the duties imposed on clients.

Notification of projects to the HSE is not required if the client is a domestic client i.e. does not carry out a business or undertaking from the premises.

1.20 What duties remain in respect of CDM when work is done for a domestic Client?

Contractors have to comply with all the duties placed on them as laid out in Part 4 and Schedule 2 of CDM 2007.

Designers have duties under Regulation 11 i.e. they must have adequate regard to health and safety when preparing their designs and must provide health and safety information to the domestic Client and contractor as necessary to ensure that health and safety standards are met.

1.21 How do the CDM Regulations apply to Developers?

When a project is carried out for a domestic Client and that person enters into an agreement with a person who carries on a business, trade or undertaking (whether for profit or not) in connection with which:

- Land or an interest in the land is granted or transferred to the domestic Client
- Construction work will be carried out on the land, and
- Following the construction work, the land will include premises which will be occupied as a residence.

then that person arranging for the construction work to be carried out will be a developer under the CDM Regulations.

All of the CDM Regulations 2007 will apply to the project.

1.22 What constitutes a "developer" under CDM?

A developer is someone who carries on a trade, business or undertaking (whether for profit or not) in connection with which:

(a) Land or an interest in land is granted or transferred to the Client; and
(b) The developer undertakes that construction work will be carried out.

In effect, a "developer" is a commercial developer who sells domestic premises before a project is complete and arranges for construction work to be carried out.

Developers in this category include Housing Associations, local councils, self-build companies and other such bodies, whether they are profit making or not national house builders.

1.23 What are the consequences of failing to comply with the CDM Regulations?

All "duty holders" have legal responsibilities under the CDM Regulations.

Health and Safety legislation is triable "either way" i.e. either on summary conviction or on indictment. This means prosecutions can be heard in the Magistrates Court or the Crown Court.

If prosecutions are brought in the Magistrates Court for contraventions of CDM Regulations then the maximum fine per offence is £5,000. If prosecutions are brought in the Crown Court then fines are unlimited and custodial sentences are possible.

Any of the following can be prosecuted if they fail to discharge their legal duties:

(1) The Client
(2) The CDM Co-ordinator
(3) The Principal Contractor
(4) The Designers
(5) The Contractors
(6) The Workers.

Since 1995 the HSE have brought 267 successful prosecutions against duty holders under the CDM Regulations 1994.

Of those 267 cases, 144 have been brought against clients, with the following contraventions:

- 20% failed to appoint a Planning Supervisor
- 20% failed to ensure that a Construction Phase Health & Safety Plan was in place
- 20% failed to ensure that Construction Phase Health & Safety Plan was suitable and sufficient
- 10% failed to provide adequate information to either contractors or designers e.g. failed to undertake an asbestos survey.

A Planning Supervisor was fined £20,000 for failing to put information about asbestos into the Pre-Tender Health & Safety Plan.

Case Studies

Number 1

Following an investigation into two accidents, one a fatality, which happened on a construction site in London, the HSE brought a prosecution against the Principal Contractor for failing to have an adequate Construction Phase Health and Safety Plan in operation over the period of the two accidents.

The HSE also prosecuted the ground works Contractor for failing to take all reasonable steps to ensure that an excavation did not collapse accidentally and for failing to take suitable and sufficient measures to prevent vehicles overrunning the edge of an earthwork.

The Construction Phase Health and Safety Plan failed to consider the hazards and risks of the ground works did not include Risk Assessments and suitable Method Statements.

The prosecutions were heard in the Crown Court due to the severity of the accidents and the Principal Contractor was fined £20,000 plus £3,000 costs. The ground works Contractor was fined a total of £20,000 for two charges under the Construction (Health Safety and Welfare) Regulations 1996 plus £3,000 costs.

Number 2

HSE Prosecution of a Client, Designer and Planning Supervisor (under CDM 1994 Regulations).

A client, designer and a planning supervisor, as well as the contractors undertaking the work, were recently prosecuted at Preston Crown Court following the fatal accident to a scaffolder from Warrington who died after falling through a fragile roof light while working on an extension to a warehouse.

According to the HSE the following were charged:

- *** plc, the clients for the work, pleaded guilty of a charge under the Regulation 11 of the Construction (Design and Management) Regulations 1994 (CDM Regulations). He was fined £2,500 and ordered to pay £1,500 costs.
- ***, a partner in a firm of architects, the project designer, pleaded guilty to a charge under Regulation 13(2) (a) (i) of the CDM Regulations. He was fined £2,500 and ordered to pay £1,500 costs.
- ***, a partner in a firm of architects, the planning supervisor for the project, pleaded guilty to a charge under Regulation 15(1) of the CDM Regulations. He was fined £2,500 and ordered to pay 1,500 costs).
- *** Ltd, the deceased's employer, pleased guilty to a charge under Section 2(1) of the Health and Safety at Work Etc Act 1974 (HSW Act). The company was fined £35,000 and ordered to pay £20,000 costs.
- *** Ltd, the Principal Contractor for the work, pleaded guilty to a charge under Section 3(1) of the HSW Act. The company was fined £25,000 and ordered to pay £15,000 costs.

The Principal Inspector for the HSE said "This was a tragic incident in which a man lost his life in circumstances that could so easily have been prevented. The dangers presented by fragile roof materials have been well known for many years, yet were not taken into consideration in this case. Considerable time had been spent planning the job and any one of those involved had the opportunity to realize that somebody could fall through the warehouse roof. Simple steps could then have been taken to prevent this death. If this had been done, the deceased would still be alive today".

1.24 What civil liability extends to the CDM Regulations?

The CDM Regulations constitute criminal law – it is a criminal act to contravene them. Anyone who does so incurs a criminal record.

Civil law applies to situations where an individual can sue another person (or corporate body) for damages due to their negligence i.e. failing in their common law duty of care.

Being prosecuted for criminal offences under CDM does not infer an automatic right to bring civil proceedings. A successful prosecution does not necessarily imply a failure in the duty of care under civil law and a successful civil case will depend on the facts of the case.

Regulation 45 of CDM 2007 states that breaches of certain duties may confer a right to make a civil claim e.g. in instances where the site was inadequately protected from unauthorized access, duties imposed under Regulations 26 to 44 were ignored, welfare facilities not provided, etc.

1.25 Will the CDM 2007 Regulations only be enforced by the Health & Safety Executive?

No, not necessarily. Generally, most construction sites come under the jurisdiction of the HSE for health and safety enforcement and therefore CDM enforcement.

But the changes introduced in the new CDM 2007 Regulations mean that there will be opportunity for local authorities to enforce CDM in those premises for which they have statutory inspection duties e.g. retail premises, offices, entertainment venues, sports facilities, etc. The Health & Safety (Enforcing Authority) Regulations 1998 set out which enforcement body is responsible for enforcing the law in various types of workplace.

Where works are construction works and generally, where they are not notifiable, the local Environmental Health Officer could be responsible for ensuring that CDM is being followed – especially in relation to site safety matters e.g. working at heights, use of equipment, provision of welfare facilities and so on.

The HSE are responsible for enforcing health and safety – and fire safety – on construction sites and if the area where the work is being carried out can be described as "a place set aside for construction work" then the HSE will enforce the CDM Regulations.

Projects which are notifiable will be enforced by the HSE.

1.26 The CDM 1994 Regulations did not apply to some constructions works carried out in a workplace in which the local authority enforced the law. Has CDM 2007 changed this?

Yes.

CDM 2007 applies to *all* construction works and apart from domestic clients, has to be applied to all construction projects.

Previously, two sets of Regulations applied to construction works – CDM Regulations 1994 and Construction (Health, Safety & Welfare) Regulations 1996. It may have been the case that the CDM 1994 Regulations did not apply to a project but the "welfare" Regulations did, thereby ensuring that health and safety standards were set. So, even if they did not need to enforce CDM '94, officers could have enforced the Construction (Health, Safety & Welfare) Regulations 1996.

CDM 2007 Regulations combine both the former CDM 1994 Regulations and the Construction (Health, Safety & Welfare) Regulations 1996 in to one comprehensive set of Regulations. So there can therefore be no exemptions to the Regulations as this would mean some construction activity would be outside of the law.

1.27 Does CDM 2007 apply to all forms of building procurement such as PFI, PPP, etc.?

Yes – both because irrespective of how the project is funded or acquired works will be construction and design work will take place.

Project originators should take on the duties of clients and ensure that a CDM Co-ordinator is appointed early on into the project and that HSE notification is made in a timely manner.

The role of client can be assigned to other parties during the course of the project but there must always be someone clearly identified in that role.

Where multiple clients are discussing and agreeing a new project one of them must be appointed the Client in order to comply with CDM. It would be sensible to record this as part of the management process and communicate to all other parties.

1.28 If there are several clients involved in a project do they all have duties under CDM 2007?

Not necessarily.

Regulation 8 CDM 2007 allows for multiple clients to elect either one or more of them to be classed as the Client under the Regulations.

All clients have to agree in writing that such an election has taken place and once this has been done, no other client will be required to fulfil client duties, except that they will all be required to provide any information to other parties as appropriate.

If there is no agreement in writing or no one client wants to take on the full responsibility for everyone else, then all clients will remain as clients under CDM 2007 and their various CDM Co-ordinators will have to ensure good co-operation and co-ordination of information, etc. during the course of the project.

It would be sensible to start the initial project meetings with a review of CDM responsibilities and minute the discussion and decisions made regarding the appointment of duty holders, etc.

1.29 What are the general duties imposed on everyone under CDM 2007?

The CDM Regulations 2007 have emphasized the need for duty holders to be competent to do the jobs they are appointed to and for everyone involved in the construction project to be competent and/or properly supervised.

The Regulations contain general duties on:

- Competence
- Co-operation
- Co-ordination.

The Regulations also contain the general duty that all persons involved in planning, designing or carrying out construction work follow the principles of prevention in respect if health and safety.

1.30 What are the duties in respect of competence?

Regulation 4 requires that no person on whom the Regulations place a duty i.e. Clients, CDM Co-ordinators, Principal Contractors, Contractors, Designers, shall appoint or engage anyone to undertake the role of:

- CDM Co-ordinator
- Designer
- Principal Contractor
- Contractor.

unless he has taken reasonable steps to ensure that the person to be appointed is competent to carry out the tasks required.

No person who is offered an appointment under CDM 2007 should take on the role unless they themselves are satisfied that they are competent to undertake the role.

No person may instruct a worker to carry out or manage construction work unless he is competent or under the supervision of a competent person.

This means that anyone who wishes to appoint or engage someone to do a construction activity, or design a new building, etc., must be satisfied that the person they want to appoint has sufficient knowledge and experience to carry out the tasks.

Often accidents and major incidents occur because the people carrying out potentially complex tasks have been appointed because they are the cheapest and can do the job the quickest. There have been many incidents involving the collapse of buildings because the contractors involved have not understood the correct demolition sequences or have not been experienced enough in undertaking the job thereby being ignorant of the hazards and risks involved.

All persons have this duty so if for instance, a designer needs to engage the services of say, a structural engineer he must ensure that the person is competent and not just appoint the cheapest company.

A client should ultimately be satisfied that his whole design, procurement and implementation team is competent and should therefore ensure that if members of the project team appoint sub-contractors, etc. that they have rigorous procedures in place for assessing competencies.

1.31 What are the duties in respect of co-operation?

Regulation 5 CDM 2007 states that every person concerned in a project, on whom a duty is placed by the CD Regulations shall:

- Seek the co-operation of any other person concerned in any project involving construction work at the same or an adjoining site so far as is necessary or function under the Regulations; and
- Co-operate with any other person concerned in any project involving construction work at the same or an adjoining site so far as is necessary to enable that person to perform any duty under the Regulations.

Any person who is working on a project under the control of another person must inform that person if they believe there is any issue which could affect their own or others health and safety.

This Regulation requires everyone to co-operate with each other, whether part of the project team for the construction project or involved in a separate project which may have some influence on the main project.

Co-operation will be required for instance, in areas with shared access. The retail store may be having an extension built and the rear service yard may be used by other retailers. The Principal Contractor may need to co-ordinate his materials deliveries with the other premises stock deliveries. The other retailers must co-operate with the Principal Contractor so as to enable the Principal Contractor to operate safely.

HSE see the improvement in co-operation between duty holders as a significant step in improving the incidents of poor health and safety e.g. often contractors will have to handball materials long distances increasing musculoskeletal risks because others have blocked the vehicle entry to the delivery area.

1.32 What are the duties in respect of co-ordination?

All persons concerned with a project on whom duties are placed by the Regulations shall co-ordinate their activities with one another in a manner which ensures, so far as is reasonably practicable, the health and safety of persons:

- Carrying out the construction work and
- Affected by the construction work.

Co-ordination means to:
"bring together and cause to work together efficiently".

If all parties involved in a project co-operated with and co-ordinated their work with each other, better planning would result and conflicts involving health and safety would be significantly reduced.

How many times do builders work and mechanical and electrical installations clash on site, often creating unsafe working conditions when one or other of the parties has to "make do" to make something work on site, usually ending up working in an unsafe way to get the job done.

Because co-ordination of activities is seen as such a critical factor in reducing health and safety incidents, the Planning Supervisor role has been replaced by the CDM Co-ordinator.

1.33 Do the CDM Regulations 2007 apply to Scottish projects?

Yes.

The Regulations, in their entirety apply in Scotland.

1.34 Do the CDM Regulations 2007 apply in Wales?

Yes.

The Regulations, in their entirety apply in Wales.

1.35 Do the CDM Regulations 2007 apply in Northern Ireland?

Yes, although the Regulations have the addition of Northern Ireland in the title.

The Regulations, in their entirety apply in Northern Ireland.

1.36 Does CDM 2007 apply to term contracts?

CDM 2007 does not apply to the "term contract" but may apply to the individual projects undertaken as part of the term contract.

HSE's general view regarding term contracts is that any F10 notification will be project specific within the general term of the contract.

Notification for a term contract for general work that may not take place is not of benefit.

HSE are interested in works of construction which will last more than 30 days or take more than 500-person days. These projects will be notifiable whenever they occur in a term contract.

CDM 2007 applies to all construction work so really the only consideration will be to decide whether the work package is notifiable.

It may be sensible to address the roles and responsibilities of CDM 2007 and outline the procedural approach for compliance within the term contract.

A practical approach to a term contract in which minor works are carried out to a range of buildings by the same team is to appoint someone in to a similar role to that of the CDM Co-ordinator i.e. someone to co-ordinate all the health and safety issues, any design issues across all the contractors teams.

2

Health and Safety Legislation

2.1 What legislation applies to the construction industry and its projects?

All legislation which places duties on employers and others to ensure the safety of their workers and those affected by their undertaking is relevant to the construction industry.

The following legislation can be applied to construction projects:

- Health and Safety at Work Etc Act 1974
- Health & Safety (First Aid) Regulations 1981
- Electricity at Work Regulations 1989
- Control of Noise at Work Regulations 2005
- Construction (Head Protection) Regulations 1989
- Personal Protective Equipment Regulations 1992
- Manual Handling Operations Regulations 1992
- Reporting of Injuries, Diseases and Dangerous Occurrences Regulations 1995
- Control of Vibration at Work Regulations 2005
- The Health & Safety Signs and Signals Regulations 1996
- The Confined Spaces Regulations 1997
- The Provision and Use of Work Equipment Regulations 1998
- The Lifting Operations and Lifting Equipment Regulations 1998
- The Management of Health & Safety at Work Regulations 1999
- The Regulatory Reform (Fire Safety) Order 2005
- Control of Asbestos Regulations 2006
- Control of Lead at Work Regulations 2002
- Control of Substances Hazardous to Health Regulations (COSHH) 2002
- Working at Height Regulations 2005
- Corporate Manslaughter and Corporate Homicide Act 2007 (to be implemented April 2008).

In addition to the above list, one of the key pieces of legislation for construction safety is:

The Construction (Design and Management) Regulations 2007.

2.2 What are the main duties covered by the Health and Safety at Work Etc Act 1974?

The Health and Safety at Work Etc Act 1974 is the "underpinning" legislation which governs virtually all other health and safety law.

The Act sets out the general parameters of what is expected of employers and other persons in respect of ensuring their health, safety and welfare.

Regulations are subsidiary legislation made under the enabling powers of the Health and Safety at Work Etc Act 1974. Contravening Regulations is an offence and prosecutions can be brought regarding each offence. In addition, there may be a breach of the more general principles of health and safety enshrined in the 1974 Act and additional charges could be brought under various sections.

The main sections of the 1974 Act are as follows.

2.2.1 Section 2   General duty of employers to their employees

It shall be the duty of every employer to ensure so far as is reasonably practicable, the health, safety and welfare at work of all his employees.

.... the matters to which that duty extends include:

(1) Provision and maintenance of plant and equipment, and systems of work that are safe and without risks to health.
(2) Arrangements for ensuring safety and absence of risks to health in connection with the use, handling, storage and transport of articles and substances.
(3) The provision of such information, instruction and training as is necessary to ensure the health and safety at work of his employees.
(4) The maintenance of any place of work under the employers control in a condition which is safe and without risk to health and the provision of means of access and egress that are safe.
(5) The provision and maintenance of a working environment that is safe, without risks to health and adequate with regards facilities and arrangements for their welfare at work.
(6) Provision of a written statement of his policy in respect of health and safety of his employers.

2.2.2 Section 3

It shall be the duty of every employer, to conduct his undertaking in such a way as to ensure, so far as is reasonably practicable, that persons not in his employment are not exposed to risks to their health and safety.

2.2.3 Section 4

It shall be the duty of each person who has control of premises to take such measures to ensure that premises, means of access to and egress from available for use by persons using the premises are safe and without risks to health.

2.2.4 Section 7

It shall be the duty of every employee while at work :

(1) To take reasonable care for the health and safety of himself and of other persons who may be affected by his acts or omissions at work; and
(2) To co-operate with his employer so far as is necessary so as to enable the employer to undertake statutory duties.

2.2.5 Offences, penalties and prosecutions

Offences

These include the following:
(a) Failing to comply with the general duties on employers, employees, the self-employed, persons in control of premises, manufacturers, etc.
(b) Failing to comply with any requirement imposed by Regulations made under the Act.
(c) Obstructing or failing to comply with any requirements imposed by inspectors in the exercise of their powers.
(d) Failing to comply with an Improvement or Prohibition Notice.
(b) Failing to supply information as required by a Notice served by the Health & Safety Commission (e.g. investigations into major accidents).
(b) Failing to comply with a Court Order to remedy the cause of an offence.

Penalties

Most offences under the Health & Safety law are "triable either way" i.e. summarily or on indictment.

Breaches of employers duties under Section 2 Health and Safety at Work Etc Act 1974 carry a maximum fine of £20,000 *per offence* if tried summarily.

Breaches of employers duties under the numerous Regulations enacted under the Health and Safety at Work Etc Act 1974 carry fines of up to £5,000 per offence if tried summarily. In some circumstances, fines are up to £20,000 per offence.

Where the case is heard in the Crown Court, fines are unlimited.

Breaches of Improvement or Prohibition Notices carry either 6 months imprisonment if heard in the Magistrates Court or up to 2 years imprisonment if heard at Crown Court.

Prosecutions

(a) Offences by Companies, Corporate Bodies and Directors (Health and Safety at Work Etc Act 1974 Section 37)

The Health and Safety statutes place duties upon limited companies and/or functional directors.

Where an offence is committed by a body corporate, senior persons in the hierarchy of the Company may be *individually liable*.

If the offence was committed with the consent or connivance of, or was attributable to any neglect on the part of the following persons, that person himself is guilty of an offence and liable to be prosecuted:

- Any functional director
- Manager
- Secretary (Company)
- Other similar officer of the company
- Anyone purporting to act as the above.

The conditions for liability under Section 37 are:

- Did the person act as the Company?
- If he acted in that capacity, did he act with neglect?

Directors, Managers and Company Secretaries are personally liable for ensuring that Corporate Safety duties are performed throughout the Company. They may be able to delegate the specific responsibilities but that does not absolve them of liability.

(b) Offences due to the act of "another person"

Section 36 of Health and Safety at Work Etc Act 1974 states that where an offence is due to the act or default of another person, then:

That other person is guilty of the offence, and

A second person (e.g. body corporate) can be charged and convicted whether or not proceedings are taken against the first-mentioned person.

Case Studies

Fatal injuries to conveyor belt worker

A 29-year-old man died after becoming trapped in a slew conveyor pit he was trying to clear.

The machine had no guard and no emergency stop button within reach.

Safety standards at the company were virtually non-existent and contraventions were found of the working at height regulations, transport safety, use of chemicals, equipment safety and the provision of welfare facilities.

An Area Manager with the Company and the Managing Director were both sentenced to imprisonment – 9 months and 12 months respectively.

Charges were brought under Section 2, Health & Safety at Work Etc Act 1974.

Construction worker dies after being hit by a dumper truck

A construction worker dies after being struck by a dumper truck when the brakes failed and it rolled out of control down a slope.

The prosecution involved both the contractors and individuals as the investigation identified an unsafe site.

The worker who dies was driving the dumper truck but he was not qualified or trained to do so.

The construction company was fined £75,000 plus costs of over £50,000 for breach of both Section 3 Health & Safety at Work Etc Act 1974 and a regulation in the 1994 CDM Regulations.

The sub-contractor was fined £100,000, with costs of £17,643, for breaches of Section 3 Health & Safety at Work Etc Act 1974.

Individual directors were fined £2,500 and £1,000 for breaches of Section 7 Health & Safety at Work Etc Act 1974.

2.3 What are the main duties contained in the Management of Health and Safety at Work Regulations 1999?

Employers must make suitable and sufficient assessment of the risks to health and safety of their employees and to non-employees affected by their work. Each risk assessment must identify the measures necessary to comply with relevant statutory provisions.

Risk assessments must be in writing where there are more than five employees and they must be reviewed regularly.

Employers must introduce appropriate arrangements for effective planning, organization, control, monitoring and review of the preventative and protective measures. These arrangements must be in writing where there are five or more employees.

Where appropriate, employees must have health surveillance if they are exposed to hazards which could affect their health and safety and well being.

Employers must establish and effect appropriate procedures to deal with emergencies e.g. evacuation, major chemical spillage, explosion and so on. Employees must stop work immediately and proceed to a place of safety, if exposed to serious imminent and unavoidable danger.

Employers must provide comprehensive and understandable information to all employees on risks identified in the risk assessments, emergency procedures, preventative and protective measures, competent personnel.

Employees must receive appropriate health and safety training on recruitment and throughout their employment.

Employers must appoint a "competent person" to assist in undertaking the measures necessary to comply with statutory provisions. Competent persons can be employees or external advisors/consultants are permissible if there are no suitable employees.

Employers must consider the health and safety of young persons at work, pregnant women and nursing mothers.

Contact must be made by employers with external services e.g. fire, police, emergency services so that any necessary measure can be taken in the event of an emergency or rescue.

Multi-occupied sites must have a plan for co-operation and co-ordination with one employer taking the lead role in respect of health and safety.

Employers must take into account the capabilities and training of all employees before assigning them tasks, etc.

Employees are under a duty to use equipment, materials, etc. provided to him by his employer in accordance with safe systems of work, any training, etc.

Employees must inform their employer of any matter relating to their own or others health and safety.

Temporary workers are to be afforded health and safety protection, information on hazards risks, health surveillance as appropriate, etc.

2.4 What are the key things I need to know about the Health and Safety (First Aid) Regulations 1981?

Every employer must provide equipment and facilities which are adequate and appropriate in the circumstances for administering first aid to employees.

Employers must make an assessment to determine the needs of his workplace. First aid precautions will depend on the type of work and therefore the risk, being carried out.

Employers should consider the need for first aid rooms, employees working away from the premises, employees of more than one employer working together, non-employees.

Once an assessment is made the employer can work out the number of first aid kits necessary by referring to the Approved Code of Practice.

Employers must ensure that adequate numbers of "suitable persons" are provided to administer first aid. A "suitable person" is someone trained in first aid to an appropriate standard.

In appropriate circumstances the employer can appoint an "appointed person" instead of a first aider. This person will take charge of any situation e.g. call an ambulance and should be able to administer emergency first aid.

Employers must inform all employees of their first aid arrangements and identify trained personnel.

2.5 What are the main duties in the Control of Noise at Work Regulations 2005?

Employers must carry out Noise Assessments where employees are likely to be exposed to:

- Noise above 80 dB(A), known as the first action level
- Noise above 85 dB(A) pascals, known as the second action level.

Competent persons must carry out the Assessments and identify employees exposed to the noise.

Noise Assessments must be reviewed on a regular basis and whenever circumstances change.

Records must be kept of all Noise Assessments.

Employers have a duty to reduce the noise to which employees are exposed, if it measures 85 dB(A) or above.

Ear protection zones must be designated where appropriate.

Information, instruction and training must be given to all employees.

2.6 What are the main provisions of the Electricity at Work Regulations 1989?

- All systems, plant and equipment to be designed to ensure maximum practical level of safety.
- Installation and maintenance to reflect specific safety requirements.
- Access, Light and Working Space to be adequate.
- Means of cutting off power and isolating equipment to be available.
- Precautions to be taken to prevent charging.
- No live working unless absolutely essential.
- Specific precautions to be taken where live working is essential.
- All persons to be effectively trained and supervised.
- Responsibility for observing safety policy to be clearly defined.
- All equipment and tools to be appropriate for safe working.
- All personnel working on electrical systems to be technically competent and have sufficient experience.
- Work activity shall be carried out so as not to give rise to danger.
- Electrical system must be constructed and maintained to prevent danger. "Danger" means the risk of injury.

2.7 What are the key things I need to know about the Manual Handling Operations Regulations 1992?

Employers must, as far as is reasonably practicable, avoid the need for employees to undertake and manual handling operations at work which involve risk of injury.

Employers must carry out risk assessments of all manual handling operations, which cannot be avoided. Assessments must be in writing.

Employers must take appropriate steps, following assessments, to reduce the risk of injury, to the lowest level reasonably practicable.

Employers must take appropriate steps to provide employees, undertaking manual handling operations, with general indications and/or precise information on weight and size. of each load and handling activity undertaken.

Employers must review assessments regularly, especially if there is any cause to believe that there has been any significant change in handling operation.

Employees must make full and proper use of any system of work provided by the employer concerning steps to reduce risks.

Consideration must be given to repetitive actions involving small weights.

2.8 What are the main duties I need to know about in the Personal Protective Equipment Regulations 1992?

Employers must ensure that suitable personal protective equipment (PPE) is provided to employees who may be exposed to risks to their health and safety, except where the risk has been controlled adequately by other means.

PPE must be suitable and appropriate to the risks and work place conditions. It must suit the worker due to wear it and afford adequate protection.

If more than one sort of PPE must be worn, they shall be compatible.

PPE must be assessed to ensure that it is suitable for the tasks. Any assessments shall be reviewed regularly.

PPE must be maintained in an efficient state, in efficient working order and in good repair.

Where PPE is needed, suitable accommodation for it must be provided.

Adequate and appropriate instruction and training must be given to employees in the use of PPE.

Employees must use PPE provided, return it to its accommodation after use and report loss or defects.

Employers must consider PPE as a "last resort" and must address all other risk reduction methods.

2.9 What are the main duties in the Reporting of Injuries, Diseases & Dangerous Occurrences Regulations 1995?

Where any person dies or suffers any of the injuries or conditions specified in Appendix 1 of the Regulations, or where there is a "dangerous occurrence" as specified in Appendix 2 of the Regulations, as a result of work activities, the "responsible person" must notify the relevant enforcing authority.

Notification must be by telephone or Fax and confirmed in writing within 7 days.

Where any person suffers an injury not specified in the above-mentioned appendices but which results in an absence from work of more than 3 calendar days the "responsible person" must notify the enforcing authority in writing.

The "responsible person" may be the employer, the self-employed, someone in control of the premises where work is carried out or someone who provided training for employment.

Where death of any person results within 1 year of any notifiable work accident the employer must inform the relevant enforcing authority.

When reporting injuries, diseases or dangerous occurrences the approved forms must be used either F2508 or F2508A.

Records of all injuries, diseases and occurrences which require reporting must be kept for at least 3 years from the date they were made.

Accidents to members of the public which result in them being taken to hospital as a result of the work activity must be reported.

Incidents of violence to employees which result in injury or absence from work must be reported.

2.10 What are the key things I need to know about the Lifting Operation & Lifting Equipment Regulations 1998?

Lifting equipment and operations are covered by the Regulations, including lifts, hoists, eye bolts, chains, slings, etc.

Lifting equipment must be adequate in strength and stability for each load.

All equipment used for lifting persons must be safe so that they cannot be crushed, trapped, struck by or fall from the lifting carrier.

Where safety ropes and chains are used they must have a safety co-efficient of at least twice that required for general lifting operations.

Lifting equipment must be installed or positioned in such a way that it reduces the risk of the equipment striking a person or of the load drifting, falling freely or being released unintentionally.

Suitable devices must be available to prevent persons from falling down a shaft or hoist way.

Equipment used for lifting must be marked with safe working loads, including that used for lifting people.

If equipment is not suitable for lifting persons it must be marked accordingly.

Lifting operations must be properly planned, appropriately supervised and carried out in a safe manner.

Lifting equipment must be regularly inspected and tested and a report of its condition produced:

(i)   Before being used for the first time.

(ii)   After assembly and being put into service for the first time at a new site.

Lifting equipment used for lifting *people* must be examined and tested every 6 months.

All other lifting equipment e.g. goods hoists must be examined and tested every 12 months, unless the competent person deems them to need inspecting more frequently.

Records must be kept although these can be electronic.

Persons carrying out examination and testing must be competent.

2.11 What are the key provisions of the Provision and Use of Work Equipment Regulations 1998?

Employers must ensure that work equipment is suitable by design, construction or adaptation for the work for which it is provided.

Employers must take into account any risks in the location where the equipment will be used e.g. wet areas and electrical equipment.

Work equipment must be maintained in a suitable condition and in good working repair. Where there is a maintenance log it must be kept up-to-date.

Equipment will have to be guarded where necessary, be able to be isolated from power sources, be used in the right environment, have display warnings, etc.

Employees must be given adequate health and safety information on the use of equipment, the dangers, etc. They must also be given suitable training.

Persons must be nominated to carry out maintenance and repairs on equipment, must be suitably trained and competent.

Work equipment must be capable of isolation from any power source.

Maintenance operations are to be carried out when equipment is stopped, unless work can be done without exposure to health and safety risks.

Warning notices must be displayed as necessary adjacent to equipment.

Work equipment includes installations.

Mobile work equipment is included, as are woodworking machines and power presses.

Second hand equipment brought into a business after 1 December 1998 will be classed as new equipment.

2.12 What are the key things I need to know about the Control of Substances Hazardous to Health Regulations 2002?

Employers must carry out assessments of all work activities involving the use of substances hazardous to health which might pose health risks to employees or others.

All assessments must be suitable and sufficient and must be in writing.

Exposure of employees to hazardous substances must be prevented or otherwise controlled – the hierarchy of Risk Control must be followed.

PPE is only permitted when other control measures are not practicable.

Control measures must be maintained in efficient working order and good repair. Ventilation equipment must be checked annually.

Monitoring of exposure of employees to substances must be carried out where appropriate.

Employees must be given health surveillance where appropriate and records kept for 40 years.

All employees exposed to hazardous substances must receive adequate information, instruction and training on the health risks created by their exposure to substances.

Information must be made available from manufacturers. Hazardous substances with Occupational Exposure Limit (OEL) or Maximum Exposure Limits (MEL) levels must be used in strict compliance with Codes of Practice, etc.

2.13 What are the main provisions of the Regulatory Reform (Fire Safety) Order 2005?

Requires all responsible persons to assess fire hazards within their workplace i.e. produce Risk Assessments.

Applies to all employers and persons in control of premises, including Principal Contractor, Clients, Landlords and others.

Appropriate controls must be implemented to either eliminate or control the risks.

Suitable provision must be made for fire fighting equipment.

All fire safety equipment must be maintained and tested regularly.

Emergency plans must be in place for raising the alarm in the event of a fire.

Staff must receive adequate training in fire safety matters.

There must be a fire detection system available which is capable of warning all persons of the risk of fire.

Suitable means of escape in case of fire must be provided from all areas of the building.

Fire Risk Assessment, procedures, etc. must be regularly monitored and updated when circumstances require.

2.14 What do the Control of Vibration at Work Regulations 2005 require employers to do?

The Control of Vibration at Work Regulations require you to:

• Assess the vibration risk to your employees.

- Decide if they are likely to be exposed above the daily exposure action value (EAV) and if they are:
  - Introduce a programme of controls to as low a level as is reasonably practicable.
  - Provide health surveillance (regular health checks) to those employees who continue to be regularly exposed above the action value or otherwise continue to be at risk.
- Decide if they are likely to be exposed above the daily exposure limit value (ELV) and if they are:
  - Take immediate action to reduce their exposure below the limit value.
- Provide information and training to employees on health risks and the actions you are taking to control those risks.
- Consult your trade union safety representative or employee representative on your proposals to control risk and to provide health surveillance.
- Keep a record of your risk assessment and control actions.
- Keep health records for employees under health surveillance.
- Review and update your risk assessment regularly.

2.15 What are the main provisions of the Confined Spaces Regulations 1997?

The definition of confined space covers, amongst others, the following:

- Trenches
- Vats
- Silos
- Pits
- Chambers
- Sewers
- Vaults
- Wells
- Internal rooms.

Employers have a duty to ensure that their employees comply with the Regulations, and that they, as employers ensure that employees are not exposed to risks to their health, safety and welfare.

Any environment which could give rise to "specified risks" can be covered by the Regulations. Specified Risk includes:

- Injury from fire or explosion
- Loss of consciousness through a rise in body temperature or asphyxiation
- Drowning
- Injury from free flowing solids causing asphyxiation
- Anything which prevents an escape from space.

Risk Assessments are required for work in confined spaces.
Wherever practicable, work in any confined space shall be avoided.

Entry into a confined space is prohibited unless suitable rescue arrangements have been put in place.

Workers and rescuers must be trained in the hazards and risks associated with confined spaces.

2.16 What are the key things I need to know about the Control of Lead at Work Regulations 2002?

The Regulations apply to any type of work activity which is liable to expose employees and any other person to lead.

Lead is any lead, including lead alkyls, lead alloys, any compounds of lead and lead as a constituent of any substance, which is liable to be inhaled, ingested or otherwise absorbed by persons. Lead given off in exhaust fumes from road traffic vehicles is excluded.

Employers must:

- Assess the risk to health from lead
- Carry out Risk Assessments
- Identify and implement measures to prevent or adequately control exposure to lead
- Record the significant findings of the assessment
- Protect employees where exposure is deemed to be significant
- Issue employees with protective clothing
- Monitor lead in air concentrations
- Place employees under medical surveillance
- Ensure high standards of personal hygiene
- Provide employees with information, instruction and training
- Identify the contents of containers and pipes
- Prepare procedures for emergencies, accidents and incidents.

2.17 What are the key things I need to know about the Control of Asbestos Regulations 2006?

Employers are responsible for the health and safety of all persons, whether at work or not.

Employers are not responsible for providing information, instruction and training to persons who are *not* his employees unless those persons are on the premises where the work is to be carried out.

A duty holder must manage any risks from asbestos and must arrange for assessments to be carried out to determine whether asbestos is, or is likely to be present, on the premises.

A duty holder is anyone who has a responsibility or contract for maintenance or repair of the premises, or who is or has control of any part of the premises.

More than one duty holder can exist for a premises and the level of responsibility will be determined by how much they have to contribute to repairs and maintenance.

Premises have to be inspected to determine whether asbestos is present.

Risk assessments must be made.

A premises register is developed and maintained as to the whereabouts of asbestos, how it is managed, etc. The register must be kept up-to-date and available to anyone who may need to see it.

No employer shall carry out work on asbestos until its type has been identified.

No employer shall carry out work on asbestos unless a risk assessment has been carried out.

Findings shall be recorded.

Asbestos types must be identified.

Control measures need to be assessed.

A suitable written plan must be produced for any work with asbestos.

The Plan shall be kept on the premises for as long as the work continues.

Notification must be made at least 14 days before works are proposed to commence.

Every employer has to ensure that adequate information, instruction and training is provided for employees.

Information on the risk assessment must be given, including risks to health, precautions to be taken, control and action levels, etc.

Information, instruction and training must be given at regular intervals.

Employers must ensure that exposure to asbestos is prevented and where this is not possible, reduced to the lowest practical level.

The number of employees so exposed must be as low as possible.

Substances may be used as a substitution for asbestos if they pose lesser risks.

Any control measures must be used correctly.

All control measures must be properly maintained, repaired, etc.

Ventilation equipment must be thoroughly tested and examined.

Suitable records of tests, maintenance, etc. shall be kept for at least 5 years.

Adequate and suitable protective clothing must be provided for employees working with asbestos.

Clothing so exposed to asbestos must be disposed of or cleaned.

Clothing must be removed from site in sealed containers.

If any personal clothing is exposed to asbestos it shall be treated as protective clothing or either disposed of or cleaned.

Emergency procedures shall be developed and tested e.g. safety drills.

Information on emergency arrangements shall be available.

Suitable warning and communication systems shall be available.

Any unplanned release of asbestos shall be dealt with immediately so as to mitigate the effects, restore the situation to normal and inform persons who may be affected.

Every employer shall prevent or reduce to the lowest level reasonably practicable, the spread of asbestos from any place where work is carried out under his control.

Premises and plant shall be kept clean.

Premises, or parts, shall be cleaned after works are completed.

All employers shall ensure that any area where asbestos work is being undertaken is designated an "asbestos area", or a "respirator zone" where the level of asbestos exceeds the control level.

Notices designating these areas must be displayed.

All employees, other than those needing to work in the area, shall be excluded from the area.

No eating, drinking to smoking shall take place in a designated area, and other places for these activities must be available.

Asbestos fibres in the air shall be monitored by employers at regular intervals and when changes occur that may affect exposure.

Air monitoring is not required if exposure will be below the action level.

Records of air monitoring shall be kept.

Records shall be kept for 5 years or 40 years if they form part of a health record.

Records should be available to those who wish to see them.

Employers must ensure health records are kept of employees exposed to asbestos unless exposure is less than the action level.

Records shall be kept for 40 years.

Medical surveillance shall be given every 2 years to employees exposed to asbestos.

Adequate washing and changing facilities must be provided to those employees exposed to asbestos.

All asbestos must be stored or transported in sealed containers properly labelled.

No person shall supply products containing asbestos.

- An employer may exercise the defence of "due diligence" or that he took all reasonable precautions to avoid the commission of the offence.

2.18 What duties do employers have for the health and safety of non-employees?

Under Health and Safety Legislation, employers are responsible for the health and safety of persons who are not their employees or not in their employ.

The Courts consider the employment status of self-employed and casual/contract labour differently to that of the Inland Revenue. A person can be self-employed for tax reasons, but employed for health and safety reasons especially where the employer instructs such persons as part of their business.

Three key legal cases confirm employers responsibilities for others:

(i)   R versus Associated Octel (Criminal Court)
(ii) Lane versus Shire Roofing Company (Oxford) Ltd (Civil Court)
(iii) Nelhams versus Sandells Maintenance Limited and Another (Civil Court).

2.18.1 Associated Octel

Operated a chemical plant and instructed a specialist contractor to maintain and repair the plant during its annual shut down.

During a repair operation, a light bulb operated by one of the contractor's employees burst, lighting the cleaning fluid he was using and badly burning him.

Both the contractor and Octel (the Client) were prosecuted, Octel for failing to ensure the safety of persons other than its employees.

Found guilty. Octel appealed to the Court of Appeal.

Court of Appeal agreed initial Crown Court decision. Agreed that an employers "undertaking" can include the activities of a third party over which the employer had no control. The employer must show that it has done enough to protect the third party's safety.

The Court decided that an employer, by instructing a contractor to perform, would be "conducting his undertaking" for the purposes of Section 3 Health and Safety at Work Etc Act 1974, and as such has duties to ensure the safety of that contractor's employees.

2.18.2 Shire Roofing Company (Oxford) Ltd

Lane was a self-employed (for tax purposes) roofing contractor who was hired by Shire Roofing to re-roof a porch at a private house.

Lane told Shire that he could do the job for £200 "all in". To make the job profitable he needed to use a ladder and not a scaffold or tower scaffold which was offered by the roofing company.

Shire Roofing did not get involved with Lane or the job, did not supervise him, etc.

During the course of his work, have slipped and fell and suffered severed head injuries.

Lane brought a civil action for damages against Shire, claiming they had breached a common law duty of care as an employer to ensure his health and safety as an employee.

The Court found that Mr Lane was independent contractor and not entitled to any common law duty by another.

He appealed to the Court of Appeal (Civil Division).

The Court of Appeal reversed the decision and stated that Lane had been hired as a labourer rather than as an independent contractor and that his tax status was irrelevant for health and safety purposes.

The Court of Appeal reviewed the "control" test which the earlier Court had applied.

The control test obliges the Court to ask questions about the relationship between the parties, such as who decides what activities are to be performed and how they are to be carried out, who allocates the resources for the job and who sets the timetable, etc.

The Court of Appeal decided that the first "control test" was inconclusive as it had failed to consider the professional skills and discretion that Lane had in making decisions about the work. When determining "employment" to correct question was not "who was in control", but rather "whose business was it".

Shire Roofing's supervisor felt that it was his responsibility to provide safety aids, materials and plant, etc. to short term labourers, such as Lane.

The Court of Appeal decided that Shire Roofing had a duty of care to Lane and awarded him £102,000 damages, plus costs.

2.18.3 Sandells Maintenance Ltd

Nelhams was a permanent employee of Sandells.

Sandells "loaned" Nelhams to another company – Gillespie, telling him he was under the "complete control" of Gillespie.

Gillespie asked Nelhams to do some painting. He was told he would have to do the job from a ladder as scaffolding could not be used. Nelhams asked for the ladder to be footed but was told that no-one was available. Whilst climbing the ladder, it slipped and Nelhams fell.

Civil claims were brought against Sandells and Gillespie by Nelhams. At trial, only Gillespie was found guilty as Sandells had handed over control of Nelhams.

Gillespie appealed to the Court of Appeal.

The Court of Appeal found that both employers were equally responsible for Nelhams' safety.

Even though there was an arrangement and "loan" of an employee from one employer to another, the primary employer remains liable for the safety of his employees, despite the fact that he might loan them to another.

"The general employer cannot escape liability if the duty has been delegated and then not properly performed" – stated by the Court of Appeal.

Mr Nelhams was awarded damages, and both employers were found responsible.

But, the Court of Appeal then went on to rule that only Gillespie was wholly responsible for the accident as it failed to find labour to foot the ladder. Gillespie should therefore pay all the damages and should indemnify Sandells completely thereof.

2.19 What legal precedence has been set by the Octel, Shire and Sandells cases?

An employer can now be prosecuted for failing to secure the safety of an independent contractor's employees, whether or not he has been involved with that contractor's activities.

An employer can be held liable for the safety of a person holding themselves out to be independent and self-employed if they instruct such persons as part of their business.

A temporary employer can be held financially liable for the injuries suffered by the employee of another.

3

Risk Assessments

3.1 What is hazard and risk in relation to health and safety?

Hazard means the potential to cause harm.

Risk means the likelihood that harm will occur.

A typical *hazard* on a construction site is falling from heights. The *risk* of falling is dependant on what control measures have been implemented e.g.:

- The risk of falling is less if guard rails are provided
- The risk of falling is less if safety harnesses are used
- The risk of falling is less if safe means of access is provided – stairs and not ladders, tower scaffolds and not ladders, etc.
- The risk of falling is less if materials do not need to be carried
- The risk of falling is less if wind and weather conditions are considered
- The risk of falling is less if adequate working space is provided e.g. not perched on a ledge.

The *hazard* of falling from heights will be high, but the *risk* of falling will be low if guardrails are used.

3.2 What are Method Statements?

Method Statements are written procedures which outline how a job is to be done so as to ensure the safety of everyone involved with the job, including persons who are in the vicinity.

A Method Statement equates to a "safe system of work" which is required under the Health and Safety at Work Etc Act 1974.

A Method Statement can be used as the control measure needed to eliminate or minimize the risks involved in carrying out the job.

Method statements are required for all demolition work.

3.3 Who must provide Risk Assessments?

A Principal Contractor should receive risk assessments from all the other Contractors, Sub-Contractors and self-employed tradesmen working on the site. These will detail what the hazards associated with their tasks are e.g. noise from drilling equipment and will include details of how the risks from the hazards e.g. noise induced hearing loss, can be eliminated or reduced.

When each of the Contractor/Sub-Contractor risk assessments have been reviewed, the Principal Contractor must consider whether he needs to do anything else to protect other workers in the area, i.e. forbid certain work activities in certain areas, etc. If so, he will need to do an additional Risk Assessment which identifies how as Principal Contractor he is going to manage and control the combined risks of several Contractors.

Risk Assessments need only identify *significant* risks involved in carrying out a work activity. Routine risks and every day risks such as crossing the road to get to the employee car park need not be included.

Where anything unusual or uncommon is to be undertaken on the site, a risk assessment will be essential. Where works involve significant hazards e.g. working in confined spaces, working at heights, working with harmful substances then risk assessments are legally required and the control measures identified could be incorporated into a Method Statement which operatives are required to follow.

3.4 What responsibilities do employers have in respect of completing Risk Assessments?

The Management of Health and Safety at Work Regulations 1999 require employers to complete Risk Assessments for all activities undertaken at work for which there is a risk of injury or ill-health.

The Regulations place an absolute duty on employers to undertake suitable and sufficient assessments of:

- The risks to health and safety of employees to which they are exposed whilst at work; and
- The risks to health and safety of persons not in their employment arising out of or in connection with the conduct by them of their undertaking. The key details are:
  - Complete risk assessments
  - Ensure that they are suitable and sufficient.

Any significant findings must be put in writing.

Employees must be given information, instruction and training on the Risk Assessments so that they understand the hazards and risks of the tasks to be undertaken. Employees must also know what control measures the employer has determined as being appropriate to minimize the risks of injury from the hazards identified.

3.5 The law refers to suitable and sufficient Risk Assessment? What does this term mean?

A Risk Assessment should:

- Identify the significant risks arising out of the work
- Enable the employer to identify and prioritize the measures needed to be taken to ensure that all relevant statutory provisions are complied with
- Be appropriate to the nature of the work
- Remain in force for the duration of the work
- Be regularly reviewed.

A Risk Assessment does not need to be perfect but must be "the best which can be produced given the knowledge available at the time". It must be "proportionate" to the risks associated with the tasks. It must therefore be suitable and sufficient for the hazards and risks which need to be addressed.

3.6 Who is best placed to carry out a Risk Assessment?

There are no legally required qualifications for carrying out Risk Assessments. The Management of Health and Safety at Work Regulations 1999 requires that whoever carries out a Risk Assessment must be "competent" to do so.

Competency is not defined in any of the Regulations but guidance which accompanies various Regulations refers to:

"appropriate knowledge, training and experience".

The person carrying out the Risk Assessment must be able to:

- Identify hazards
- Judge the consequences of the hazards i.e. likely injury
- Determine how likely it is that the harm will be realized
- Identify what control measures are needed to reduce the risks
- Determine whether there is imminent risk of serious personal injury and prohibit the continuation of the task.

It would be sensible for the person carrying out the risk assessment to have had some training in the techniques and processes involved in conducting the assessment.

Site Agents should be trained in the methodology of Risk Assessments.

Site Agents should enquire about the competency of any sub-contractors who purport to be experienced in carrying out Risk Assessments.

Under the Construction (Design and Management) Regulations 2007 any person who appoints a contractor must assess their competency and resources in respect of health and safety.

3.7 What are the five steps to Risk Assessment?

One of the best and most simple of guides to understanding Risk Assessment is the Health and Safety Executive's (HSE's) free leaflet:

"Five Steps to Risk Assessment" : INDG 163

This very useful guide lists the five steps as:

Step 1   Look for hazards
Step 2   Decide who might be harmed and how
Step 3   Evaluate the risks and decide whether existing precautions are adequate or more should be done
Step 4   Record the findings
Step 5   Review and revise the Assessment if necessary.

3.7.1 Step 1: Hazard Spotting

Walk around the site and look for anything which could cause harm. Ignore the trivial things and concentrate on those hazards which could cause serious harm.
  Site hazards will include:

- Trip hazards
- Working at heights
- Using electrical equipment
- Manual handling
- Noise
- Harmful substances
- Unguarded drops, voids
- Easy access to site for unauthorized persons
- Dust
- Low temperatures
- Vehicles and mobile plant
- Insufficiently trained operatives
- Using lifting equipment
- Falling objects
- Unstable structures
- Tower scaffolds
- General scaffolding
- Inadequate working platforms
- Exposure to live services
- Use of welding equipment
- Likelihood of fire.

3.7.2 Step 2: Decide who might be harmed and how

Consider anyone who could be exposed to the hazard and who could be harmed as a result. Injury can be minor or major – it really should not matter as good health and safety is about preventing injury and ill-health – no matter how insignificant.

Consider in particular, people unfamiliar with the site, namely:

- Visitors
- Delivery drivers
- Clients
- Short duration visiting contractors.

Also consider:

- Young workers
- Trainees and apprentices
- Trespassers to the site
- Members of the public
- Anyone working on site, or visiting, with a disability
- Women who may be pregnant (not an unusual event as more women are training in the building trades).

3.7.3 Step 3: Evaluate the risks and decide whether existing precautions are adequate

Having identified the hazards on site and who might be harmed, review what steps are currently in place to reduce the risk of injury. Usually, something is in place either consciously or not e.g. 110 V tools are used to reduce the risk of electric shock/electrocution from using 230 V tools.

Are the steps that are in place enough to prevent someone being injured at all?

These steps are called *control measures*.

If you think more could be done, then the Risk Assessment process requires you to implement additional controls where you can.

For example, you have identified a ladder as having the potential to cause harm – people could fall off. Their injuries could be severe. The current "control" is to tie the ladder at the top to stop it moving. You have reduced the risk of people falling off the ladder due to the ladder moving unexpectedly.

But, you could do more to reduce the risk of falling from ladders altogether by requiring all operatives to use a mobile tower scaffold. This is a safer means of access than a ladder.

But, mobile tower scaffolds create their own hazards and so you will need to control these.

Have you done, are you doing, everything that is "reasonably practicable" to keep the construction site safe?

Draw up an Action List of the controls which you could put in place. Consider the following:

- Eliminating the hazard completely
- Trying a less risky option
- Preventing access to the hazard e.g. guarding

- Remove people from the hazard
- Issue personal protective equipment (PPE)
- Provide methods with which to deal with the hazard immediately e.g. washing facilities to reduce skin contamination from substances.

3.7.4 Step 4: Record findings

It is always sensible to keep a record of what steps have been taken to address hazards and risks. If employers have five employees or more, Risk Assessments must be in writing, although you have only to record "significant" risks.

This means writing down the significant hazards and your conclusions of who might be harmed, how and how often and what control measures are in place, or *are* to be put in place to control the risks.

These written records are often referred to as "Risk Assessments". They do *not* need to be rocket science and they do *not* need to be perfect. Risk Assessments must be "suitable and sufficient".

You need to be able to show:

- That you reviewed hazards
- That you thought about what could happen, why, how and when
- That you considered who could be harmed and why
- That you considered all *known* steps that were practicable, which you could introduce or follow to reduce the hazards
- That the residual or remaining risk was lower than when you started
- That you informed people of the hazards and risks and what safety steps they should follow to reduce the hazards to acceptable levels
- That you recognize that the hazard and risk situation needs to be reviewed from time to time.

If one of the Statutory Enforcement Agencies visit site and note hazards, they will want to see, almost without fail, your Risk Assessments.

Keep Risk Assessments until at least the end of the project. If anyone has had an accident of any significance, keep the appropriate Risk Assessment for at least 6 years – remember they can launch a civil claim in the courts many years after they had the accident.

It is acceptable to refer to any other health and safety records, procedures, manuals, etc. in your Risk Assessment.

Risk Assessments do not have to be in any specific format, but must cover the information required.

3.7.5 Step 5: Review and revise the Assessment

Site conditions change frequently and the controls which were in place last week may no longer be appropriate to new working areas, procedures, etc.

Review the Risk Assessment:

- Has anything changed on site
- Has the weather affected anything

- Have new operatives started work
- Have new openings been made or new work areas created
- Has new plant been brought to site
- Has site layout changed.

Check that the precautions you have introduced to reduce risk are working effectively. If you are having a number of accidents occurring on the site it could indicate that controls are not working. Review and investigate all accidents and then review Risk Assessments.

Case Study

The Principal Contractor was responsible for ensuring the delivery of materials to site. The delivery area incorporated the rear access road which was shared by a neighbouring retail premises. There were hazards to both the site operatives and adjoining tenants from the delivery vehicles and the off-loading of materials. Hazards included moving vehicles, restricted access to the roadway for emergency vehicles, off-loading materials from the lorries, dust, noise, falling objects. The risks from the hazards included being knocked over, being hit by materials, noise induced hearing loss, breathing in dust and exhaust fumes, etc.

The Principal Contractor formulated the risk assessment, identifying the above as the hazards and risks and determining the control measures needed to eliminate or minimize the risks. These included having a banksman to guide in the delivery vehicles, setting specific delivery times, liaising with the adjoining tenants, providing lifting devices, requiring engines to be switched off during delivery, avoiding reversing vehicles wherever possible, etc.

The Principal Contractor then prepared a short Method Statement which was given to the Site Foreman to follow when deliveries occurred.

The preparation of this Risk Assessment and Method Statement was the Principal Contractor's responsibility because he had overall management control of these activities and could co-ordinate everyone else's deliveries to site.

3.8 What are "site-specific" Risk Assessments?

HSE Inspectors prefer to see site specific risk assessments on every construction site.

There are Risk Assessments which relate to the actual site conditions and the actual type of construction project and are not general or "generic" Risk Assessments which address all sorts of hazards not really relevant to the site in question.

The value of Risk Assessments is that they consider the hazards and risks of an activity which employees, or others, are undertaking and they refer to actual site or workplace conditions which actually *exist*.

Many construction contractors have manuals of "generic" Risk Assessments and often these are referred to as the site Risk Assessments. They are not.

The "generic" Risk Assessments can be used as the basis of the Risk Assessment, but the competent person must still review the actual site conditions to see if the hazards identified on Risk Assessment are less or more likely to cause harm due to the local circumstances. For example, the hazards of moving vehicles on a construction site are quite common and can be generically assessed *but* on your site, the risks of injury may be increased because of "blind corners". The generic Risk Assessment may not address the hazard of "blind corners". The HSE Inspector will expect the "site-specific" Risk Assessment to do so.

3.9 Does one Risk Assessment satisfy all the Regulations?

Risk Assessments for various work activities are required under the following Regulations:

- Management of Health and Safety at Work Regulations 1999
- Manual Handling Operations Regulations 1992
- Personal Protective Equipment at Work Regulations 1992
- Health and Safety (Display Screen Equipment) Regulations 1992
- Control of Noise at Work Regulations 2005
- Control of Substances Hazardous to Health Regulations 2002
- Control of Lead at Work Regulations 2002
- Control of Asbestos at Work Regulations 2006
- Dangerous Substances and Explosive Atmospheres Regulations 2002.

The requirements of the Management of Health and Safety at Work Regulations 1999 are the over-riding superior Regulations and they are super-imposed over all the other Regulations.

A thorough Risk Assessment process under the Management of Health and Safety at Work Regulations 1999 will probably satisfy the requirements for all other Regulations, but other Regulations may contain specific control measures which will need to be considered e.g. exposure to hazardous substances under Control of Substances Hazardous to Health Regulations 2002 (COSHH) and whether health surveillance is needed.

When considering the hazards of working in confined spaces, the requirements of the Confined Space Regulations 1997 need to be considered even though the Regulations themselves do not require specific Risk Assessments.

4

The Client

4.1 Who is a Client?

A Client is any person or company who is involved in a business, trade or undertaking for whom a construction project is carried out.

To fall into the definition of Client, it is not necessary for the construction works to actually take place, but the term refers to any person or company who first thinks about having a structure built, repaired, refurbished, demolished or maintained.

Concept and feasibility schemes are commissioned by a "Client" under Construction (Design and Management) (CDM) and even though actual construction work may not take place, the principles of CDM must be applied.

Clients can include:

- Businesses
- School Governors
- Charities
- Insurance companies
- PFI consortia
- Developers
- Social clubs or voluntary organizations
- Private members clubs
- Individuals who are undertaking a business e.g. pub landlords, restauranteurs.

4.2. Why do CDM Regulations have to apply to Clients?

Research has shown the Health and Safety Executive (HSE) over the years that Clients have tremendous influence on the planning, organization and implementation of a construction project. If the Client has a cavalier attitude to health and safety then it is likely that everyone on the project will develop a similar approach. If standards of health and safety on construction sites are to be improved, Clients must "lead from the top" and must ensure that adequate provision, including time and financial resources, is made for health and safety on every project.

Clients also often hold considerable amounts of information about their building or the project to be undertaken and by ensuring that this information is made available, designers and contractors can plan for any known hazards, etc. Hence the duty placed on clients to provide information.

4.3 Does the Client's Agent post still exist under CDM 2007?

No.

The 2007 Regulations have revoked the position of Client's Agent for all projects which came into being after 6 April 2007.

Research by HSE found that many clients thought they could abdicate their responsibilities for CDM 1994 by appointing a Client's Agent thus contributing to the poor performance of the Regulations.

HSE expects clients to be fully committed to their duties under CDM 2007 and they have clearly stated that those responsibilities cannot be delegated.

Where a project already has a Client's Agent appointed i.e. on projects started before April 2007, the position can remain in place until either the end of the project or a period of five years elapses from the coming into force of the Regulations, whichever is the soonest.

A Client's Agent will have to agree in writing to continue in the role of Client's Agent as they will be responsible for the Client's duties under CDM 2007 and not CDM 1994.

4.4 What are the duties of a Client under CDM Regulations?

The following duties must be carried out for every project to which CDM applies:

- Will not be able to appoint Clients Agent.
- Must ensure that suitable arrangements are made to manage the project safely.
- Appoint a CDM Co-ordinator and Principal Contractor for all notifiable construction projects.
- Ensure that suitable welfare arrangements are in place during construction.
- Ensure that designers and contractors are promptly supplied with information relevant to their purposes.
- Ensure that contractors (Principal Contractor on notifiable projects) are informed of the minimum time to be allowed for planning and preparation before construction commences.
- Ensure that construction work does not start on notifiable projects before a Construction Phase Health & Safety Plan is in place.

- Ensure that suitable arrangements are made to protect the health and safety of users of any structure designed as a workplace, as well as of construction workers, cleaners and maintenance workers.

4.5 What are a Client's specific duties regarding the arrangements for managing construction projects?

Every client has to take reasonable steps to ensure that arrangements are in place, or made, for managing a project, including the allocation of sufficient time and resources, by persons with a duty under the CDM Regulations and that they are suitable to ensure that:

- The construction work can be carried out so far as is reasonably practicable without risk to the health and safety of any persons.
- The requirements of Schedule 2 of the Regulations are complied with in respect of any person carrying out construction work.
- Any structure designed to be used as a workplace has been designed taking into account the provisions of the Workplace (Health, Safety and Welfare) Regulations 1992 which relate to the design of, and materials used in, the structure.

The Client should also take reasonable steps to ensure that the arrangements made are maintained and reviewed throughout the project.

The Client is expected to ensure that things are done, not necessarily to do them, himself.

Reasonable steps have to be taken to review the procedures and arrangements adopted by all duty holders, not just contractors.

Clients will need to ensure that all duty holders have good health and safety knowledge, have access to competent advice, know what their legal responsibilities are, have suitable trained and experienced people available to work on the project, etc.

Some of the arrangements will be covered in any general competency assessments e.g. health and safety arrangements, accident records and so on, and these may not need to be reproduced for every project.

CDM 2007 Regulations contain the requirement in Regulation 9, for clients to be more involved in ensuring that proper planning and management of the project has taken place.

It is important that there is proper clarity of roles and responsibilities for the project. If the project is notifiable, these roles and responsibilities can be included in the Pre-Construction Information Pack.

Clients must influence the time available to the specific stages of the project and whilst many clients are sceptical that "if you give a contractor an inch, they will take a mile" and therefore prolong the construction phase unduly, they must now demonstrate that they have taken the time needed for the project and must show that they have not enforced unreasonable timescales.

One of the greatest concerns of both major and smaller contractors is that clients impose unrealistic timescales on them, forcing them to accelerate works and take shortcuts thus compromising health and safety on site.

Designers often complain that clients force them to cut costs and omit preferred safety solutions e.g. roof-edge protection, thereby not allowing them adequate financial resources to design the project safely.

Clients must ask the project team how they are proposing to communicate, co-operate and co-ordinate the information for the project and will need to be satisfied how the project meetings, design meetings or progress meetings are being scheduled.

The circulation of meeting minutes may be reasonable evidence that the project is being managed effectively.

4.6 Can another party act on behalf of the Client in ensuring that management arrangements are in place?

Yes, although the Client cannot delegate their legal responsibilities for ensuring that the arrangements are in place.

If the project is notifiable i.e. will last more than 30 days or take more than 500-person days, the Client can ask the CDM Co-ordinator to act on their behalf and assess the management arrangements in place.

Where projects are smaller and not notifiable, the Client can ask the competent person appointed to advise them on health and safety – as required under the Management of Health & Safety at Work Regulations 1999, to advise and assist them.

Clients are not expected to be experts in managing construction projects, but they are expected to understand that they cannot ignore the requirement for good management and health and safety on a project. They are expected to ask relevant questions and to query less than satisfactory answers.

Clients are expected to make a judgement call on the appropriateness and adequacy of management arrangements – doing nothing is not an option.

4.7 What are a Client's duties regarding the provision of welfare facilities on a construction site?

A Client has to ensure that provision has been made for adequate welfare facilities prior to the commencement of construction works.

The Client does *not* have to provide welfare facilities but if they have them available e.g. a hotel is being refurbished and the existing staff toilets can be used, they are expected to co-operate with the contractor and make them available if at all possible.

The Client must assist the contractor in achieving the provision of a good standard of welfare facilities, especially sanitary accommodation. If, for instance, the Client can facilitate early connections into mains drainage, this would enable the contractor to install flushing toilets thus avoiding the use of inferior chemical toilets.

A Client does not have to tell the contractor what facilities to provide as the contractor is responsible for ensuring that adequate facilities are provided and need not stipulate numbers of WC's, urinals or wash hand basins, etc. However, a Client should be able to assess general adequacy of the facilities and should encourage the contractor to provide flushable WC's, running water and mains drainage, or drainage to suitable tanks.

Chemical closets, inadequate water supplies, etc. are not really acceptable and generally imply to the workforce that tier welfare is not important.

4.8 Does information only have to be provided by the client for notifiable projects?

No.

The duty to provide information is placed on the Client for *all* projects.

If the project is notifiable, the CDM Co-ordinator can assist and advise the Client about what information needs to be provided and to whom.

But, if the project is small and not notifiable, then the Client must ensure that the information is provided.

The best was is to ask the designers and other members of the team, and the contractor, if appointed early enough, what information they would hope to have and what information they believe to be essential in order for them to do their job safely.

The information need only be proportionate to the job in hand and does not need to be reams of paper and manuals.

The arrangements for managing the project could set out what information is available from whom and list who is responsible for providing it and when.

4.9 What information must be provided to the CDM Co-ordinator?

In order to comply with the Client's duties under CDM, the CDM Co-ordinator must be provided with all relevant information in order for them to carry out their duties (Regulation 15 CDM 2007).

Relevant information will enable the CDM Co-ordinator to ensure that the Construction Information Pack is prepared, that the Designers are advised of key information to help them complete their design risk assessments and integrate health and safety into the project.

Examples of information a Client is expected to be able to provide:

• Site details	—	(1) Address of site (2) Owner of site if different from Client (3) Nature of project (4) Commencement and duration of works, including time between appointment of contractor and start on site
• Appointments made by Client	—	(1) Architects (2) Quantity Surveyors (3) Structural Engineers (4) Specialist Designers (5) Project Managers (6) Principal Contractor (if known)
• Management arrangements	—	(1) Roles and responsibilities (2) Communication procedures (3) Site security issues (4) Welfare provisions (5) Health and safety objectives (6) Times for planning and preparation
• Location of site	—	(1) Environmental information (2) Proximity to roads, houses, schools, etc. (3) Proximity to rivers, overhead power cables, power sub stations, etc.
• Site boundaries	—	(1) Demarcation/demise area (2) Fencing and boundaries (3) Ownership of adjoining land/premises (4) Details of "communal" shared area
• Condition of structure	—	(1) Description of building (2) State of dilapidation (3) Structural information (4) Fragile materials
• Ground conditions	—	(1) Ground type (2) Ground stability (3) Contamination (4) Presence of underground tanks, channels, pits, etc.

(*Continued*)

• Location of mains services	–	(1) Gas
		(2) Electrical/power
		(3) Telephone
		(4) Communications
		(5) Fuel supplies
		(6) Overhead power lines
• Hazardous substances	–	(1) Asbestos
		(2) Lead paint
		(3) Dust/fumes
		(4) Pigeon/rodent contamination
		(5) Chemical contamination
		(6) Oils, petrol
• Fire Safety Arrangements		
• Overlap with Clients Undertaking	–	(1) Employees still at work
		• Any hazardous processes
		• Hazardous areas
		• Traffic routes
		• Site safety rules
• Existing health and safety file or other building information e.g. DDA access statements, fire safety strategies, risk assessments and so on		

4.10 Do detailed surveys need to be commissioned in order to provide this information?

The more information available at the planning stage of the project regarding potential health and safety issues, the more likely it will be that the project can be programmed efficiently and safely.

Finding out about the presence of hazardous substances, e.g. asbestos once the construction phase has started means downtime, costs and potential health risks to all operatives and others. Breaches of legislation are inevitable.

It makes sound business sense to obtain as much information about the site before progressing the design phase. Asking a Structural Engineer to advise on load-bearing walls and the necessary temporary works will prevent a major on-site collapse, which could cause major and fatal injuries to those on site.

Commissioning the following would not be unreasonable:

- Building survey including services
- Structural survey
- Asbestos and other hazardous substance survey

- Pest survey
- Contaminated land survey
- Environmental noise survey
- Land use survey, including adjacent premises, etc.
- Overhead and underground power lines/cables survey.

The CDM Regulations are about reducing hazards and risks to health and safety i.e. being prepared to deal with what is there because of being forewarned.

If asbestos is present, it poses a *serious* health hazard. The HSE expects all Clients to have identified the existence of asbestos on site *before* construction works start and to have included proposals to deal with it in the Pre-Construction Health & Safety Pack e.g. remove it completely or encapsulate it, label it and manage the risks. (Preferred option is always to remove asbestos from site following the legal requirements of the Control of Asbestos Regulations 2006 unless considered more hazardous.)

Each project needs to be assessed on its merits – if the building is a new "developers' shell" then detailed surveys will not be necessary as the developer will have dealt with these issues. If the building is post-1980, it is unlikely (though not guaranteed) that it will have sprayed asbestos coating and lagging as insulation and fire protection material but it may have asbestos boards or asbestos cement tiles.

A high street retail premises which has been in existence for decades may not need a contaminated land survey unless a review of previous planning permissions indicates an industrial factory or petrol station had been present on the site and extensive ground works are proposed.

4.11 When does a CDM Co-ordinator need to be appointed and what are they supposed to do?

The CDM Regulations 2007 create a new role of CDM Co-ordinator to help co-ordinate the health and safety aspects of a project and to undertake specific duties. They must be appointed on all notifiable projects.

A CDM Co-ordinator can be an individual or a number of individuals e.g. a Company with health and safety experience and knowledge, construction and design knowledge and experience of project planning.

A good CDM Co-ordinator will ensure that duties as a Client under the Regulations are being met, that Designers fulfil their responsibilities and that the Principal Contractor manages site-safety issues.

Effective co-ordination and planning will mean a more efficient project with reduced hazards and risks, better safety management, less lost time for accidents and a greater efficiency in respect of resources and time management.

A good CDM Co-ordinator helps to ensure that health and safety is thought out in advance and that others involved in the project are directed or inspected in respect of their knowledge, approach and implementation of health and safety issues.

CDM Co-ordinators need to be competent and have adequate resources to do the job. They must have a comprehensive knowledge of the construction process, and health and safety. They should not make the project more difficult or complicated. They should not create mounds of paperwork. They should not be obstructive.

They should be helpful, give advice, co-ordinate, encourage, question, compromise, cajole and communicate with all members of the design team on *any* issue relative to health and safety of the project.

4.12 What are a Client's duties regarding the provision of information on a project?

Every client has a duty to provide information to designers and contractors prior to the commencement of construction which is relevant to them to enable them to carry out their duties under CDM 2007.

All available information must be collated in to the Pre-Construction Information Pack.

The amount of information provided must not only be relevant to the project in hand but also include everything which could be considered relevant, such as

- Information of the site, existing building
- Any survey information
- Future use of the building
- In-house health and safety procedures.

Clients perhaps unwittingly withhold information about a site or a building and designers and contractors can be exposed to unnecessary risks to their health and safety. Typical examples would be not advising the project team and contractors about asbestos-containing materials, or fragile roof surfaces, rotten flooring, overhead power lines, etc.

4.13 What are the differences between the Planning Supervisor role found in CDM 1994 and the new CDM Co-ordinator role found in CDM 2007?

The duties of the CDM Co-ordinator are broadly similar to those of the former Planning Supervisor, but the approach is expected to be very different.

The CDM Co-ordinator is seen as the facilitator that ensures good co-operation, co-ordination and communication amongst the project team, predominantly in respect of health and safety matters, although their expertise could be used to help facilitate environmental and other issues.

The Client is expected to rely heavily on the CDM Co-ordinator and the role is seen as the "Client's Critical Friend" or Client's Best Friend.

The CDM Co-ordinator must advise the Client on how to meet their duties under CDM 2007 and assist them in doing so.

The former Planning Supervisor role was largely an ineffective role as they merely had to ensure that things were done with very few duties to actually do anything themselves.

The CDM Co-ordinator's role is more of a "doing" role.

The CDM Co-ordinator provides the client with a project advisor on health and safety management and ensuring effective planning of the work, together with the provision of advice on assessing the competency and resources of duty holders.

Under the former 1994 CDM Regulations, the Planning Supervisor had to ensure that the Health and Safety File was compiled. Now, under CDM 2007 they must prepare it for the Client.

The relationship between Client and CDM Co-ordinator will be critical – a mutual trust and respect has to be established and the CDM Co-ordinator must be able to voice any concerns regarding health and safety matters on the project without fear that they will be dismissed from the project.

Competent, professional, ethical CDM Co-ordinators may decide that there are some Clients for whom they do not wish to work.

4.14 What are the duties of a Client before giving the go ahead for a notifiable construction project to start on site?

You must ensure that you have complied with the CDM requirements to appoint *competent* designers, a CDM Co-ordinator and Principal Contractor. Each of these functions must also have adequate resources to carry out the tasks required.

An assessment of competency and resources needs to have been carried out and this process should be on-going throughout the duration of the project, although the Client has no legal duty to monitor the standards of his appointees during the project.

The only specific legal duty on the Client prior to construction work commencing on a notifiable project is to ensure that the Construction Phase Health and Safety plan has been prepared and that welfare facilities will be provided.

The Client does *not* have to approve the Plan but must ensure that it complies with Regulation 23 and he must be satisfied that Regulation 22 (provision of welfare facilities) will be complied with during the construction phase.

If necessary, the Client can ask the CDM Co-ordinator for advice on the suitability of the Plan and welfare facilities.

The Client also has a duty to ensure that the Principal Contractor has sufficient *time* to prepare the Construction Phase Health and Safety Plan prior to the commencement of works and he must provide information to the Principal Contractor on the amount of time he has for preparation and planning for the project.

4.15 As Client, I have changed my mind about the design of the structure before starting on site. What should I do?

The Design Team is responsible for ensuring that the Client is kept informed of CDM responsibilities and they must advise you that they will need to reconsider your new design and produce new Design Risk Assessments. These Design Risk Assessments will need to be handed to the CDM Co-ordinator so that they can ensure that key information of health and safety hazards and risks is passed to the Principal Contractor. In turn, the Principal Contractor must ensure that he addresses the design risks within the Construction Phase Health and Safety Plan.

If, as Client, you have prepared the designs you will be a designer under CDM 2007 and the design risk information may need to be provided by your own organization.

Again, the Principal Contractor may need time to alter the Plan and make additional safety arrangements, so as Client, you must ensure that you give the Principal Contractor adequate time to comply with the regulations. Delaying the date when construction work is intended to start will be almost unavoidable.

HSE Inspectors would look very closely at any unrealistic timescales for the project if there was evidence that designs had kept changing prior to commencement of works.

4.16 When must the Client appoint a Principal Contractor?

A Client must appoint a Principal Contractor as soon as is practicable after the Client has such information about the project that it will be notifiable to the HSE and the construction work involved in it as will enable the Client to comply with the requirements to assess the competency and resources of any Contractor appointed to carry out or manage construction works.

It is essential that the Principal Contractor is appointed in sufficient time to develop the Health and Safety Plan before construction works start.

If a contract is to be negotiated with a preferred Contractor it is essential that they are formally advised that they are deemed to be Principal Contractor for the project under CDM Regulations. A formal letter of appointment would be sensible. An example is shown in Appendix 13A.

If a construction contract is to be awarded by competitive tender, it would be useful to include an introductory paragraph in the Pre-Construction Information Pack, which is sent to all tenderers, that states that the success-ful tenderer for the project will be deemed to be the Principal Contractor as defined in the CDM Regulations.

Information and/or clarification on who will be appointed Principal Contractor can be included in the Bill of Quantities or Employers Requirements.

4.17 What does the Principal Contractor have to do and will I have to pay additional fees?

The Principal Contractor is responsible for site safety, assessing competency and resources of Contractors, co-ordinating work activities, organizing site training, setting up communication between contractors, monitoring site safety.

The Principal Contractor must develop any pre-construction health and safety information into the *Construction* Phase Health and Safety Plan. This plan is a critical document which sets out the rules for safety on site, how the site will be managed, arrangements for training and communication, welfare facilities, etc.

In addition, the Principal Contractor must appoint competent Contractors, carry out induction training, provide protective equipment for general use, carry out general Risk Assessments and co-ordinate the health and safety of all trades working on the project.

The Principal Contractor's duties involve co-ordinating and planning site safety. Therefore it takes time and resources, and these must be paid for. If the project is tendered on a Bill of Quantities or is negotiated then an allowance should be included for undertaking the role of Principal Contractor.

The costs will depend on the value and duration of the project.

You must, as Client, allow your Principal Contractor resources to do the job. This means that ensuring adequate financial provision is made to do the job properly. You may need to allow for a full-time Safety Manager on site – this should be included in the Pre-Construction Information Pack (if appro-priate) and an item included in the Preliminaries of any Bills of Quantities for costing purposes. Also, if a full-time Safety Manager is required, this could be included in the Employer's requirements on how the project is to be managed.

If the Principal Contractor is effective at managing, co-ordinating, planning and monitoring site safety, it should have beneficial effects on the project in both financial and timescale terms. Good site organization could reduce site accidents and down time thereby allowing the project to progress in accordance with the construction programme.

Poorly managed sites create "accidents waiting to happen". Accidents cost all parties money to a greater or lesser extent, including the Client. Therefore, additional fees in the "prelims" may be recouped many times over by having fewer accidents or near misses on site.

4.18 As Client, how do I assess the adequacy of the Construction Phase Health and Safety Plan for notifiable projects?

The Client cannot allow construction works to commence without a Health and Safety Plan having been prepared which complies with Regulation 22.

Often the Client requires the CDM Co-ordinator to advise on the suitability of the Construction Health and Safety Plan, indeed the CDM Co-ordinator legally has to be in a position to give adequate advice.

The Construction Phase Health and Safety Plan should be a practical document which is site specific. That means it must address health and safety issues relevant to the development project and the actual site conditions.

It is sensible to have a checklist of requirements which the Health and Safety Plan must meet and to compare the information given in the Plan to "model answers".

The Health and Safety Plan must be compiled by the Principal Contractor prior to commencement of construction works and it will help them if you were to require the CDM Co-ordinator to issue a guide as to how you expect the Construction Phase Plan to be laid out and what information you expect it to contain.

Construction Phase Health and Safety Plan formats can vary, but the HSE will expect it to be succinct and relevant to the project. It should not contain information on situations or hazards unlikely to be encountered on the project.

The critical question is whether the Plan is site specific i.e. does it include information about this particular project in respect of:

- Management of health and safety on the site and who is responsible for what.
- Risks involved in the construction works on *this* project.
- Any activities which need to be undertaken on the site by any person which could affect the health, safety and welfare of others at work on the site, or resorting to the site.
- Welfare arrangements for the site.
- Monitoring and reviewing health and safety issues on the site.

It is not necessary for a Construction Phase Safety Plan to be voluminous and to include every Method Statement or Risk Assessment ever produced.

It is more important that everyone on the site knows who is who, who will undertake Risk Assessments, who will organize and inspect communal access equipment (e.g. scaffolding), what training is to be provided, when inductions will be carried out, how they should raise health and safety issues on site, etc.

The Construction Phase Health and Safety Plan does not have to be totally complete before works start on site. It must include as much information as possible, particularly about site set-up procedures, initial enabling or stripping out works. The Plan needs to be developed as works progress – it is a "dynamic" document which should be updated as works progress on site.

The Plan must include details regarding welfare facilities and note how these may need to be changed as works progress.

Not all of the design issues may have been decided and Designers' Risk Assessments may be produced during the project. These need to be interpreted by the Principal Contractor and incorporated into the Construction Safety Plan. Therefore the document can never be totally complete on day one.

Once assessed, it would be wise for the Client or the CDM Co-ordinator to issue an acceptance in writing and confirmation that works can commence on site. If the Plan is not acceptable, steps should be taken to work with the Principal Contractor to improve it so that it meets the requirements of Regulation 22.

Case Study

An HSE Inspector visited a site where asbestos insulation boarding was being removed. He reviewed the Construction Phase Health and Safety Plan and determined that it was not adequate in so far as it did not comply with Regulation 22. It was too general and not site-specific and did not address the risk of asbestos-containing materials.

The Inspector served an Improvement Notice on the Principal Contractor, requiring a suitable Health and Safety Plan be developed.

A Prohibition Notice was served requiring works with asbestos insulating board to cease immediately until proper health and safety procedures had been adopted and the requirements of the Control of Asbestos Regulations 2006 complied with.

The HSE Inspector also asked the Client for the project, and the CDM Co-ordinator, to come in to the office for "discussions" regarding their competency and knowledge of the CDM Regulations and the systems they had in place for reviewing the Construction Phase Health and Safety Plan.

4.19 If awarded the construction contract on a notifiable project, is a Contractor automatically appointed Principal Contractor?

Not necessarily so. You must ensure that you properly appoint the Principal Contractor as the Client, as this will clarify responsibilities.

The Pre-Construction Information Pack may include a statement to the effect that the successful tenderer will be appointed Principal Contractor and must allow in his tender for such responsibilities.

A section may be included in the preliminaries to the Bill of Quantities or Specification for Works.

The Client cannot proceed with a construction project (for a notifiable project) without appointing a Principal Contractor. Conflict and, more importantly, legal contraventions of the Regulations could occur if the Client thinks he has appointed a Principal Contractor but the successful tenderer has not assumed the responsibility.

Obtain appointments in writing and seek clear guidelines of roles and responsibilities.

4.20 Can a Contractor refuse to accept the position of Principal Contractor for a project?

Yes, but you may call in to question their competencies for undertaking the project.

The appointment of Principal Contractor to a project is a legal duty which the Client has to fulfil. The Principal Contractor he chooses must operate a business/trade/profession as a Contractor and must be competent and have the resources to do the job. The Principal Contractor does *not* have to be the main or biggest Contractor on site but this is obviously preferable, as they will have more involvement with the site works.

If you refuse to accept the position of Principal Contractor you may be excluded from being awarded the contract. However, if you feel that you do not have the competency or resources to undertake the role, discuss this with the Client who may decide that someone else could be appointed. The CDM Co-ordinator will be able to advise the Client accordingly.

If your refusal is due to inadequate financial recompense to fulfil the legal responsibilities discuss the matter with the Client because if the Client has not allowed for adequate resources to complete the job, he could be in breach of his legal duties under CDM.

4.21 Can there be more than one Principal Contractor on a Project to which CDM applies?

There can be only one Principal Contractor at any one time on a project, although there may be many main contractors undertaking the works.

The Principal Contractor is a specific post required under the CDM Regulations with responsibility for managing and co-ordinating all of the construction phase health and safety issues. The Principal Contractor must be given overall responsibility for co-ordinating construction phases and as this is such an important function, there can only be one such appointment.

Notwithstanding the above, at the same location there could be one or more different projects being carried out for different clients and, in these circumstances, one or more Principal Contractors could be appointed. Projects have to be distinct from each other and must not rely on one another for their viability and completion. If projects share common entrances or site access, need communal lifting equipment, share services to site, etc., then it is preferable to appoint one Principal Contractor with overall site management and co-ordinating responsibility so that these "common resources" can be managed to the benefit of the site.

Each construction project could still have a main contractor responsible to their client, but a representative should liaise with the Principal Contractor. Also, CDM Co-ordinators need to co-operate so that they can relay information to the respective contractors regarding site and design hazards and risks generated by other parts of the project.

5

The CDM Co-ordinator

5.1 What is the CDM Co-ordinator in respect of the CDM Regulations?

The CDM Co-ordinator is an appointment which the Client must make to ensure that health and safety is co-ordinated during the design and planning phase of any project which is notifiable to the HSE i.e. lasts for more than 300 days or 500-person days.

The CDM Co-ordinator need not be an individual person and the function can be discharged by several bodies if appropriate i.e. specialist input at each stage of the design and planning process and the responsibilities of the post can be shared. However, the competency requirements for the role are likely to see individuals appointed.

The CDM Co-ordinator should be considered a member of the design team as it is felt that many health and safety issues relating to construction projects, and subsequent maintenance tasks, can be "designed out" in the design and planning process. The CDM Co-ordinator can help ensure that these issues are discussed and resolved early.

The CDM Co-ordinator is expected to have influence over the design team (and Client) and to highlight health and safety issues and to initiate discussions regarding ways to implement the hierarchy of risk control or the principles of prevention.

The CDM Co-ordinator must co-ordinate information relating to health and safety amongst the design team and to facilitate the sharing of information so that a holistic approach to health and safety on the project is achieved. They must ensure good communication, co-operation and co-ordination between all duty holders.

5.2 What is the purpose of the CDM Co-ordinator?

The role of the CDM Co-ordinator is to provide the Client with a key project advisor in respect of health and safety risk management matters.

The CDM Co-ordinator's role is seen as pivotal in helping to improve the overall standards of health and safety on projects and in particular, they are expected to be able to guide, advise and influence Clients into making good, sound decisions about health and safety on projects.

Clients may not be too familiar with the design and construction process and sometimes it is easy for them to get carried along by professionals who have their own interests at heart and not those of their clients. Perhaps designers, for instance, decide to change their minds about a design solution and consider little the consequences on the overall programme, budget or buildability. Ultimately, the Client will pay for the changes but a good CDM Co-ordinator may be able to step into the debate about the need for the change and bring an independent, objective view on the health and safety consequences. The trusted advisor to the Client will then be able to offer unbiased advice on a range of issues, many of which the Client will be unfamiliar with.

The CDM Co-ordinator is seen as having a pivotal role in ensuring that designers understand their roles under CDM 2007, and follow the correct procedures and thought processes to address the hierarchy of risk control and the principles of prevention in relation to health and safety.

Often, designers have excellent design creativity but sometimes lack a practical approach to buildability and future maintenance.

It is not acceptable to leave the contractors or future occupiers with the problems of "how do I build it; how do I get at it to clean it or replace defective parts".

The CDM Co-ordinator can ask these questions very early on in the design process and hopefully, act as a catalyst to "joined up thinking" by the whole design team.

The CDM Co-ordinator is seen as the conduit down which information flows and will ensure:

"the right information to the right people at the right time" (ACOP L144 HSE).

The CDM Co-ordinator is to be the Client's "best friend" – perhaps telling them things that only best friends should!

5.3 What are the key duties of a CDM Co-ordinator?

The CDM Co-ordinator has overall responsibility for co-ordinating the health and safety aspects of the design and planning phase and for the Pre-Construction Information Pack and for producing the health and safety file.

The CDM Co-ordinator must ensure that Designers have given adequate regard to health and safety within their design, in particular that they have:

- Avoided foreseeable risks to the health and safety of any person at work carrying out construction work or cleaning work in or on the structure at any time, or of any person who may be affected by the work of such a person at work.

- Combated at source risks to the health and safety of any person at work carrying out construction work or cleaning work in or on the structure at any time, or of any person who may be affected by the work of such a person at work.
- Given priority to measures which will protect all persons at work who may carry out construction work or cleaning work at any time and all persons who may be affected by the work of such a person at work over measures which only protect each person carrying out such work.

As CDM Co-ordinator you have to ensure that the Designer has followed the "Hierarchy of Risk Control" or has implemented the "principles of prevention".

The CDM Co-ordinator must ensure that Designers co-operate with one another, as far as it is necessary for each of them to comply with the requirements of the Regulations i.e. promote good Design Risk Management.

A Health and Safety File must be produced for the project and the CDM Co-ordinator must ensure that this is done and handed over to the Client. In addition, the CDM Co-ordinator must review, attend or add to the Health and Safety File as necessary so as to ensure it contains all relevant information when handed over to the Client.

An important duty of the CDM Co-ordinator is to give adequate advice to any Client or Contractor so as to enable them to comply with their duties under the Regulations and to provide adequate advice to a Client in respect of Regulation 10 – advising on the adequacy of the Construction Phase Health and Safety Plan.

The CDM Co-ordinator has to notify the project to the relevant Health and Safety Executive Office on Form F10 or similar.

Finally, the CDM Co-ordinator is expected to advise the Client on the time and resources needed to effectively discharge their responsibilities in respect of health and safety on the project. Preparation and planning time is extremely critical to managing projects safely.

5.4 When does the CDM Co-ordinator have to notify the HSE about a construction project?

As soon as the CDM Co-ordinator is appointed to the project and determines that the project is notifiable, they should notify the relevant HSE office of the construction project. Projects are notifiable if the construction works will last more than 30 days or take more than 500-person days.

All the relevant information needed for the notification form may not be known at the time of the CDM Co-ordinator's appointment. This need not delay them from making the notification – tick the initial notification box on Form F10.

As soon as further information is available e.g. confirmed start on site date, name of Principal Contractor, tick the "additional notification" box on Form F10 and send the more detailed form to the HSE.

Make sure that either the "initial" or "additional" box is clearly indicated. The HSE use F10's to assess project complexity and may believe that a notification is late if it is unclear whether it is initial or additional information.

Case Study

A diligent CDM Co-ordinator forwarded F10 to the relevant HSE office 12 weeks before the anticipated start on site date.

Approximately 2 weeks before the start on site date the CDM Co-ordinator forwarded "additional information" to the HSE advising of the confirmed start date and number of Contractors.

The CDM Co-ordinator received a letter from the HSE alleging late notification of the project and possible prosecution should future projects be so notified.

The CDM Co-ordinator immediately contacted the Principal Inspector of the HSE office to discuss the matter and advised that the initial notification was sent 10 weeks previously. The Inspector advised that when the receipt of an F10 and the start on site date was less than 15 days apart his Department issued a "standard" advisory warning letter.

The computer was programmed to automatically generate the letter unless the F10 clearly identified that it was "additional" information.

In this instance, the HSE apologized to the CDM Co-ordinator and admitted to a clerical error.

5.5 Do exploratory works need to be notified and are they the start of construction works?

Exploratory works will be notifiable if they last 30 days or more, or involve more than 500-person days.

Exploratory works usually form part of the main project works and where they do, it is the total length of the project works which is calculated for notification purposes.

Exploratory works are included in the definition of "construction works" if they involve the following:

- Site clearance
- Investigation (but not site survey)
- Excavation
- Laying and installing the foundations of the structure.

The "construction phase" of a project commences when the construction works of a project starts.

An exploration works package can be let as a separate contract on a duration of less than 30 days (it will therefore not be notifiable) but the project will come under the requirements of the CDM Regulations and all but Part 3 (notifiable projects) will be applicable.

Site survey works are those which involve taking levels, measurements, setting out and any other visual activities which generally do not involve physical activity such as drilling bore holes, taking down fixtures, etc.

5.6 Is there any recognized qualification and set of competencies which the CDM Co-ordinator must have?

Yes. The Health and Safety Executive have specifically included guidance on the level of competency a CDM Co-ordinator must have under CDM 2007 but there are no formal qualifications.

The Client has to appoint a "competent" CDM Co-ordinator and this means someone with the necessary knowledge, experience and ability to carry out the responsibilities set out in the Regulations.

There are numerous courses now available which provide a "qualification" in CDM Co-ordination but such a qualification may not necessarily equate to competency. Did the qualification require an exam to be sat which was independently assessed or was the "qualification" produced merely on attendance at the course. Would 5 days of knowledge be sufficient to give a competent understanding of construction methods, design issues, health and safety knowledge?

A formal qualification may be helpful but is not essential when appointing a CDM Co-ordinator.

5.7 Can a Client appoint himself as CDM Co-ordinator?

Yes, provided he can demonstrate that he has the competency and resources to undertake the function.

The Client must be familiar with the requirements of Regulations 20 and 21 CDM as these lay down the duties which are applicable to the CDM Co-ordinator.

The Client must be able to demonstrate knowledge of the construction process, design function, health and safety knowledge, including fire safety, and be fully conversant with construction practices, etc.

A Client who is a company, may appoint an employee as CDM Co-ordinator under the Regulations e.g. a Facilities Manager, Project Manager. They must have the resources, including time, to do the job and must not be placed in conflicting situations which are detrimental to the health and safety of the project.

If a project is notifiable and the Client fails to formally appoint a CDM Co-ordinator, they automatically taken on that role themselves. Clients will be vulnerable to prosecution if any investigation shows that they did not have the competencies to carry out the role.

5.8 Can the Client appoint the Principal Contractor as CDM Co-ordinator?

Yes, provided the Client is satisfied that the Principal Contractor is competent and has adequate resources to do the job.

If the Principal Contractor has their own Health and Safety person then they may be in a good position to discharge the duties of CDM Co-ordinator.

The CDM Co-ordinator has to have the ability to communicate with the design team in order to discuss health and safety issues. A Contractor's Safety Officer could be an ideal person to communicate with the Architect, Building Services Consultant or Structural Engineer because they are familiar with construction processes and could advise the design team that the proposed sequence of construction or programming is not the preferred one from a health and safety point of view.

The Principal Contractor must be assessed in the same way that anyone else would be in respect of the position of CDM Co-ordinator.

5.9 Does the CDM Co-ordinator have to be independent of the other members of the design team?

No. Provided the person appointed meets the requirements for competency, a member of the project team could be appointed.

Architects, Quantity Surveyors, Project Managers, Engineers, etc. can all be appointed as CDM Co-ordinator to a project even if they are already on the design team. They may have a separate "in house" division who can carry out the function, or specialist health and safety teams.

Independent CDM Co-ordinators are available and such an appointment can bring objectivity to a project in respect of health and safety matters. It may be easier for an independent CDM Co-ordinator to question a designer's design risk assessment than for the designer to objectively do it himself. The CDM Co-ordinator should act as the go-between to all parties – Client and Designer, Client and Principal Contractor, Designer and Contractor, Designer and Designer and should ensure that there is a good flow of information regarding health and safety of the project.

5.10 How soon does the CDM Co-ordinator have to be appointed to the project?

As soon as the Client believes that a project is likely to happen e.g. at feasibility stage or when outline design/planning applications are submitted.

The HSE believe the CDM Co-ordinator to be a pivotal role in the adoption of health and safety principals throughout the project. They want to ensure that the CDM Co-ordinator has been involved in the planning stage of the project and that the appointment is not just superficial. The HSE will want to be satisfied that the CDM Co-ordinator will have had the opportunity to develop information for the Pre-Construction Information Pack prior to the tendering or negotiating phase of the contract.

The Approved Code of Practice (L144) requires the CDM Co-ordinator to be appointed as soon as is "practicable" after the Client has such information about the project and the construction work involved or initial design stage.

The CDM Co-ordinator needs to be in a position to be able to co-ordinate the health and safety aspects of the design work and advise on the suitability and compatibility of designs.

They should therefore be appointed before significant detailed design work begins.

5.11 Can the CDM Co-ordinator be changed during the project?

Yes. The CDM Co-ordinator's role is best carried out by people who have the experience and this may vary from stage to stage in a project.

The CDM Regulations allow for the CDM Co-ordinator to be "terminated", "changed" or "renewed" as necessary to ensure that those appointments remain filled at all times until the end of the construction phase.

The Client must *never* allow a construction project which is notifiable to proceed at any stage without a suitable appointment of CDM Co-ordinator. If they do, they take on the responsibility by default.

The CDM Co-ordinator for the project has to be notified on Form F10 to the HSE as soon as they are appointed. If you choose to change the CDM Co-ordinator then the *new* person must notify "revised information" to the HSE.

A project may start off at feasibility stage and the Architect or Project Manager may be appointed CDM Co-ordinator. During the first phase of the project the CDM Co-ordinator assumes responsibilities for co-ordinating the design, gathering information on the site (e.g. contaminated land surveys), advising the Client and so on. The next phase of the project may be agreed as a design and build project. The role of CDM Co-ordinator could be passed from the Architect/Project Manager to the Design and Build Contractor provided they can demonstrate competency, as they would be best placed to co-ordinate health and safety issues in the design and build phase of the project.

Case Study

The Client commissioned a feasibility and concept design project for a new block of residential flats. He appointed an independent CDM Co-ordinator to the design team as he believed that this would give an objective view to the siting of the flats, access to the site and the overall design e.g. flat roof versus pitched roof, window cleaning access and so on.

Following receipt of planning permission and lease agreements, the CDM Co-ordinator drew up an initial Pre-Construction Information Pack dealing with the concept design of the building. The Client's preferred procurement route was "Design and Build" for the detailed design and construction of the project.

The CDM Co-ordinator for the concept stage advised the Client that the Design and Build Contractor would be best placed to assume the CDM Co-ordinator's role for the remainder of the project, having satisfied himself on behalf of the Client, of the Design and Build Contractor's competency and resources.

The Design and Build Contractor was duly appointed as CDM Co-ordinator and the HSE was informed by "revised information" on Form F10.

5.12 Does a Client have to take the advice of the CDM Co-ordinator?

There is nothing in the CDM Regulations which states that the Client must take the advice of the CDM Co-ordinator. The CDM Co-ordinator must be in a position to give adequate advice to the Client to enable him to comply with the Regulations concerning competency and resources of Designers and Contractors, and in respect of the adequacy of the Construction Phase Health and Safety Plan, and also on the suitability of arrangements for managing the project.

If the CDM Co-ordinator advises the Client that a Designer or Principal Contractor or Contractor is not competent or has inadequate resources and the Client ignores such advice, the HSE would consider the Client to be in serious breach of their duties under CDM Regulation 4 because "no person" shall arrange for Designers or Contractors to prepare a design or manage construction works unless reasonably satisfied that they are competent to do so.

There is little point in appointing a CDM Co-ordinator to a project if you do not value their opinion. A CDM Co-ordinator should have their Client's interest as a priority and should offer advice in respect of the health and safety aspects of the project in an objective and unbiased way. The HSE refer to the post of CDM Co-ordinator as the "Client's friend" – implying that a good working relationship is essential.

Should a Client fail to take the advice of his CDM Co-ordinator he could be in breach of Regulation 7 Management of Health and Safety at Work Regulations 1999 by ignoring the advice of his "competent" person in relation to health and safety.

5.13 Does the CDM Co-ordinator have to visit site during the design stage?

The role of the CDM Co-ordinator is to co-ordinate information in respect of health and safety for the project, and to ensure that the Pre-Construction Information Pack is completed and available to contractors and designers as necessary.

The CDM Co-ordinator needs to understand the site, design proposals, hazards and risks of the project, and how all of these things will inter-relate to the existing environment.

CDM Co-ordinators should visit site to carry out their own survey of health and safety issues and to familiarize themselves with the area e.g. busy roads, pedestrian flow, access restrictions and so on.

There is however, nothing in the CDM Regulations which requires a CDM Co-ordinator to visit site. It could be perfectly acceptable to assimilate vital information for compiling the Pre-Construction Information Pack and for issuing advice on hazards and risks in respect of the project from information supplied by way of surveys, reports, photographs, etc.

A diligent CDM Co-ordinator will always want to visit site at the beginning of a project and they should be expected to do so.

5.14 Does the CDM Co-ordinator have to carry out site safety inspections once the project has started on site?

No. There is no duty on a CDM Co-ordinator to undertake any site safety activity in respect of construction works. Therefore you do not need to pay for this service, as it does not need to be carried out under the CDM Regulations.

You may of course, appoint the CDM Co-ordinator to carry out this duty for you in addition to their legal responsibilities. It has useful benefits for ensuring standards of safety are maintained and that the Principal Contractor complies with the Construction Safety Plan. The CDM Co-ordinator is also in a good position to give advice to the Client after site safety visits. Site visits during the construction phase do, however, allow the CDM Co-ordinator to review ongoing design issues and gather information for the Health and Safety File.

If you require the CDM Co-ordinator to assume the responsibilities of a Safety Adviser then make sure there is a specific agreement to cover the duties, and agree the fee in advance.

5.15 Does the CDM Co-ordinator have to attend every site meeting?

No. There is no requirement for a CDM Co-ordinator to do this, as there is no legal duty placed on the role to be responsible for site safety.

The CDM Co-ordinator does have a duty to ensure co-operation between Designers so far as is necessary to enable each Designer to comply with the requirements of Regulation 11 CDM and this may be best achieved by attendance at site meetings so that the CDM Co-ordinator can ensure health and safety information is freely shared. However, sitting at lengthy site meetings may be an expensive way to achieve this duty when a few regular phone calls can be undertaken to each Designer and the Principal Contractor to ensure that they have all the information they need.

It would be sensible to have the CDM Co-ordinator attend a site meeting towards the end of the project when they will be able to issue requests for information to all parties for inclusion in the Health and Safety File and check the information which had already been collected.

CDM Co-ordinators may need to attend a specific design team meeting if fundamental design changes have been made to the project as these may have health and safety consequences. The role is to help co-ordinate issues and ensure the co-operation of all duty holders.

Ensure that the CDM Co-ordinator is copied in with site meeting minutes and issue instructions to them to *read* them and take whatever appropriate action to ensure they are fully up to speed with the project.

5.16 What fee should be paid for the services of a CDM Co-ordinator?

Fees can be either on a percentage of project or contract costs or can be a "fixed fee" for the project.

Percentage costs seem to vary from 0.25% to 3% depending on project values.

Sometimes a one-off fixed fee and a percentage is charged.

A fixed fee should be based on an hourly or daily rate and the anticipated number of hours/days needed to complete the statutory duties placed on the CDM Co-ordinator.

Have a clear brief of what you want the CDM Co-ordinator to do. Base your requirements on the Regulations i.e. the duties placed on the CDM Co-ordinator. If you require *other* services e.g. site safety audits, make sure this is specified as additional services and indicate whether fees are to be included in the percentage or fixed fee or to be invoiced separately on a time charged basis.

It is easy to pay a lot of money and receive poor value for money when appointing a CDM Co-ordinator. The position should be beneficial to you and your design team and be a source of advice and information on health and safety issues.

Ask the CDM Co-ordinator the following questions in order to obtain an indication of how much time they propose to devote to the project:

(1) How do you intend to gather information regarding the project?
(2) Will you be carrying out an initial site visit?
(3) How will you co-ordinate Designers responsibilities under CDM?
(4) How do you propose to check Design Risk Assessments or other documentation?
(5) How long will it take you to prepare the Pre-Construction Information Pack, submit notification to the HSE, etc.?
(6) How do you propose to check and thereafter advise me on the adequacy of the Construction Phase Health and Safety Plan?
(7) How do you propose to advise me on competency and adequacy of resources when I appoint Designers and Contractors? What time have you allowed for this activity?
(8) How often do you intend to visit the construction site? For what purpose?
(9) How do you propose to co-ordinate the Mechanical and Electrical Designers and Contractors?
(10) How will you gather information for the Health and Safety File and how long will it take you to prepare the Health and Safety File?
(11) How will you communicate your advice to me regarding the adequacy of management arrangements for the project?

5.17 As CDM Co-ordinator, my Client wants me to give advice regarding the competency and resources of the design team. What do I need to consider?

A Client must not appoint any Designer who is not competent in respect of health and safety matters relating to his design nor must he appoint anyone who does not have the resources to comply with the requirements of the Regulations.

Many Clients will delegate the assessment of the design team to their CDM Co-ordinator. If this is the case, it is important that the CDM Co-ordinator has a written procedure in place to demonstrate the steps taken.

Step one will be to issue a Designers questionnaire which seeks general information about the Designer, Partnership, Practice or Company. Information on the types of projects they normally deal with will be relevant – a design company experienced in domestic dwellings may not necessarily be competent to design industrial buildings.

Details of individual qualifications, membership of professional bodies, etc. will be relevant.

Details of their health and safety policies, procedures and practices will be needed.

Details of how they prepare and consider Design Risk Assessments would be helpful.

Have they had experiences of the type of project before? Ask for references. Go to see similar work. Ask to see examples of Design Risk Assessments.

How do they keep their employees up-to-date about health and safety issues, design innovations, etc.? Self development through reading is fine but what about their commitment to the subject of health and safety by spending money on courses, etc.

Remember that the design team includes:

- Architects
- Quantity Surveyors
- Building Services Consultants
- Structural Engineers
- Civil Engineers
- Project Managers (if they are allowed to *specify*)
- Interior Designers
- Landscape Architects.

All of them should be subjected to competency and resources checks.

Having gathered all relevant information, devise a system where you can quantitatively score their responses to their responsibilities under the Regulations.

Information on any formal or informal action by statutory authorities e.g. HSE will be relevant, as will details of any accidents which have happened to their own personnel or on sites for which they have held some responsibility.

Nominated individuals responsible for health and safety and CDM issues should be included and examples of Safety Policy documents, CDM Procedures and Protocols, etc. should be included with any responses.

Obtaining information is relatively easy. The tricky bit is assessing it objectively so that you can conclude that they are "competent". It is easy to believe that all "professionals" are competent otherwise they would not be in business. Unfortunately this is not true!

Review the information against a "benchmark" of acceptable answers. If an answer seems inconsistent or lacks details request more information.

Just because someone is honest enough to admit to having being served with a Statutory Notice does not mean that they should automatically be excluded. Seek information about what they learned from the experience, what procedures did they review and improve. Persistent contraventions of laws indicate an unwillingness to accept responsibilities and usually implies poor standards and attitudes from Management downwards. These Designers would be best avoided.

Approvals can be given for certain types of projects, projects up to X value, etc.

Encourage the Client to develop an "Approved List" of Designers which indicates their strengths and weaknesses, approval status, etc.

Review the list annually, or more frequently if necessary.

Have they really had experience of your type of project before? Ask for references. Go to see similar work. Ask to see examples of Design Risk Assessments.

See Chapter 12 on assessing competencies.

5.18 The Client wants to pay minimum attention to CDM and to do things on a shoestring. What should I do?

Refusing to accept the commission as CDM Co-ordinator would be a good starting point. Clients who refuse to accept their legal responsibilities will probably be difficult to work with in other areas and professional integrity is as important as fee income. Even more importantly, a criminal record for failing to discharge the duties of the CDM Co-ordinator will stay on your records long after the project has finished.

The Client has to allow the CDM Co-ordinator adequate resources to do the job and has to provide information regarding the construction project. Failure to do so could be a breach of the Client's statutory duties.

The CDM Co-ordinator's role relies heavily on good communication and inter-personal skills because it is largely an "influencing" role. It would be sensible to discuss with the Client the reasons for his attitude to CDM and why he wishes to avoid his responsibilities. It may be because of fees – a question of "what do I get for my money".

A professional approach by the CDM Co-ordinator would be to explain that there may be no need to appoint a separate, independent person to the role, and that perhaps another member of the design team could fulfil the duties, saving on fee expenses.

Advise the Client what in your view needs to be undertaken in order to apply CDM to the project e.g. what information should be made available and confirm such matters in writing.

Persuade the Client of the benefits of applying CDM i.e. better planning and design reduces site accidents which in turn reduces site delays and so on.

Discuss the type of Health and Safety File which the Client would prefer – perhaps the majority of it could be collected by the Principal Contractor, reducing the time needed by the CDM Co-ordinator. The CDM Co-ordinator has to prepare the Health and Safety File but there is no reason why the information is not provided by others in a form which requires little input from the CDM Co-ordinator.

CDM is an approach to planning and managing health and safety on construction sites, and afterwards, when the building is maintained. It should be seen as an integral part of the design and construction process and need not cost considerable sums of money.

Review your own fees! Are you misleading your Client about the complexities of the CDM Co-ordinator function and demanding fees for unnecessary activities e.g. site safety visits.

5.19 The Client refuses to accept my advice as CDM Co-ordinator. What should I do?

Disagreement with the Client may damage the relationship and when and if communication has irrevocably broken down, the CDM Co-ordinator may have no choice but to resign the role.

If the CDM Co-ordinator feels that they are being prevented from carrying out their legal duties and legal compliance is in jeopardy, then a resignation of the role will be essential.

But perhaps, the situation has not got to that stage and the CDM Co-ordinator should work diplomatically to advise the Client of the reasons behind the advice he gives, the implications for complying and not complying with it, the project wide benefits, etc.

Always consider alternative approaches if possible – especially if the Client's reluctance is based on perceived costs associated with your advice.

Ultimately, the Client will be legally responsible for ensuring that they comply with the Client duties under CDM 2007 and if the CDM Co-ordinator can show that they have performed their duties diligently they may not necessarily be implicated if there was any subsequent prosecution for non-compliance with CDM.

The Management of Health and Safety at Work Regulations 1999 requires an employer to appoint a competent person to advise and assist him in undertaking the measures he needs to take to comply with the requirements and prohibitions imposed upon him by or under the relevant statutory provisions. The CDM Regulations were made under the enabling provisions of the Health and Safety at Work Etc Act 1974 and are therefore "relevant statutory provisions".

The CDM Co-ordinator could be classed as a "competent person" for giving health and safety advice and the Client would be in breach of duty if they failed to consider the advice given.

It is imperative to have written systems in place as a CDM Co-ordinator which demonstrate what advice has been given to the Client and when.

If you do resign from the project it will be important to put the reasons in writing to the Client and clearly state the termination date. Advise the Client that they will assume the CDM Co-ordinator's role until they appoint a replacement.

Provided you have exercised your statutory duties competently as required by CDM, there will be no liability to prosecution because others have failed their duties.

5.20 Can the CDM Co-ordinator be appointed by verbal agreement?

Regulation 14 CDM 2007 states that any references to appointments made for notifiable projects must be in writing.

So, although Clients can verbally agree to appoint someone to the role of CDM Co-ordinator, either they, or the CDM Co-ordinator, must issue a formal appointment letter in writing.

The letter or contract of appointment should set out as a minimum the statutory duties expected of the CDM Co-ordinator and the scope of services to be provided.

Copies should be retained by both parties.

Where the CDM Co-ordinator is appointed to numerous projects the contract or letter of appointment shall clearly state this and either list all of the projects or a general statement stating the appointment is to all projects undertaken by the Client for a specific time period.

The letter of appointment should also state under what circumstances the CDM Co-ordinator acts on behalf of the Client e.g. by signing the Client's declaration on form F10 sent to the HSE.

5.21 As CDM Co-ordinator, can I sign the F10 on behalf of the Client?

This is a slightly grey area of the law as the F10 states that the form can be signed by the Client or on his behalf, but the declaration is made that the Client is aware of his duties under the Regulations.

If the CDM Co-ordinator signs the F10 on behalf of the Client, they must make sure that they have clearly advised the Client what their duties are under the Regulations.

It would be good practice for the CDM Co-ordinator to obtain an authorization from the Client that they may sign the F10 and a confirmation that the Client is aware of his responsibilities.

The HSE expects the Client to take full responsibility for their duties under CDM 2007 and therefore expects the Client to sign the F10 as this is a clear statement that they are aware of their legal duties. If someone else signs the F10 the Client could easily avoid their duties by saying that they were unaware of them because they did not sign the F10.

5.22 What steps or procedures should the CDM Co-ordinator follow in respect of co-ordination and co-operation when appointed to a project?

Once the CDM Co-ordinator is appointed, they should immediately become part of the project team. If all of the team have worked together before this will be straightforward but where the CDM Co-ordinator is new to the team they will need to form relationships with all the key duty holders i.e. designers, the Client, contractors.

The Client should give the CDM Co-ordinator a list of the Project Team. Good relations will be achieved if the CDM Co-ordinator arranges to meet with the Project Team and establishes their roles and responsibilities on the project. A visit to appropriate offices might be beneficial, especially if the CDM Co-ordinator is to advise the Client on the competency of the project team members.

The role of any lead designer should be established.

Mechanical and electrical designers and contractors need to be included, together with engineers, specialist designers, etc.

The CDM Co-ordinator should establish whether each designer has their own procedure for identifying risks on the projects and should enquire as to how this information may be shared around the team.

The CDM Co-ordinator may decide that in the interests of co-ordination, a project specific risk register or pro forma will be used and will raise this issue with the design team.

The CDM Co-ordinator is to bring real benefits to the project team and must not be seen as someone who over complicates issues or creates mountains of paperwork.

As an example, Design Risk Assessments which are in writing are not actually required under CDM 2007 but the CDM Co-ordinator has to ensure that designers consider the hazards and risks in their designs. As long as they are satisfied that such concerns are being addressed then they need not impose requirements for meaningless Design Risk Assessments.

CDM Co-ordinators may need to attend a number of design meetings in the initial stages of the project as this is where they can bring most benefits in ensuring that all designers and other duty holders co-operate with one another. Integrating mechanical and electrical design solutions into the overall design of the project will be invaluable and hopefully, will enable hazards to be eliminated.

CDM Co-ordinators are not responsible for checking the designs and the risk assessments of designers but they are expected to raise any concerns or to give feedback if they believe health and safety is being compromised: both for the construction activity and for the future use, maintenance or cleaning of the building.

Evaluate decisions made on the principles of prevention, review the design solutions chosen, but support sensible and well-reasoned arguments which provide alternative health and safety solutions.

The CDM Co-ordinator should be able to call for a design review meeting if they believe that proper co-ordination on the project is lacking.

One way to ensure that design co-ordination has been addressed is to chair a review meeting prior to the project going out to tender or starting on site.

Ask all of the designers, and the contractors, whether they have all the information they need to progress their roles safely and ask them to sign a project declaration.

If co-operation and co-ordination is not forthcoming and once you have tried various options to improve the situation, you must advise the Client that co-operation and co-ordination is not happening and that health and safety may be at risk. The Client should then step into remind all the duty holders of both his own and their legal duties.

Once the Principal Contractor is appointed, arrange to meet with them to go through all of the information they have been given and to discuss how they will manage design changes on site, provide information for the health and safety file, etc.

5.23 The Client is looking to me, as CDM Co-ordinator, to advise him that all legal duties have been met before a project starts on site. What steps or procedures should I take?

Clients, especially those not too familiar with the construction process, will often require the CDM Co-ordinator to act as their eyes and ears and to advise on the adequacy of arrangements regarding health and safety.

A number of duties have to be met before works can start on site and the CDM Co-ordinator must be able to advise the Client that they have been fulfilled.

A simple procedure which could be adopted is the one called a "Permit to Proceed".

A specific Permit to Proceed document is designed for the project by the CDM Co-ordinator and is completed and signed, as necessary by the appropriate duty holders, before works start on site.

Such a document will be able to give a clear indication that all duties have been completed or that something is missing.

If duty holders e.g. designers and the Principal Contractor have to sign that they are satisfied with the information provided to them, then there will be clear record should disputes arise in the future about, lack of information on drawings and inadequate risk assessments.

The Client has to be satisfied that adequate arrangements have been made for welfare facilities on the project and the detail of these can be included in the Permit to Proceed.

5.24 Is there anything which a CDM Co-ordinator does not have to do under CDM 2007?

CDM Co-ordinators do not have to:

- Approve the appointment of other duty holders although they are expected to give advice to the Client on their appointments.
- Approve or check designs although they do have to be satisfied that designers have addressed the hierarchy of risk control and principles of prevention.
- Approve or supervise the Principal Contractor's construction phase health and safety plan.
- Supervise or monitor works on site.

However, any of the above tasks can be carried out by a competent CDM Co-ordinator if the Client has authorized them to do so.

- Approve Principal Contractor's risk assessments and method statements.
- Advise on the detail included in any management arrangements for the project.

CDM Co-ordinators should not take on the responsibilities of being the Principal Contractor's safety advisors although if so requested by the Client, they can offer advice and guidance to the Contractor so as to ensure the good efficiency and effectiveness of the project.

5.25 The CDM Co-ordinator has to prepare or review and update a record containing information relating to the project which is likely to be needed during any subsequent construction work. What does this mean?

The CDM Co-ordinator has to prepare, where none exists, or review and update a document known as the "Health and Safety File" for the project.

The Health and Safety File is a record of the information needed to allow future construction work, including cleaning, maintenance, alterations refurbishment and demolition to be carried out safely.

Information in the file should alert building users, owners or occupiers of any residual hazards associated with the structure and should alert any workers of any potential health and safety hazards, especially maintenance and refurbishment workers.

The Health and Safety File must be prepared by the CDM Co-ordinator. Under the previous CDM Regulations 1994, the Planning Supervisor had to ensure that the Health and Safety File was compiled. Under the 2007 Regulations, the CDM Co-ordinator must actually prepare the document.

The structure and content of the Health and Safety File should be agreed between the Client and the CDM Co-ordinator at the beginning of the project.

The CDM Co-ordinator should issue guidance to the design team and Contractors on the type and detail of information they will be requiring.

The Health and Safety File should focus on health and safety information and does not need to replicate the Building or Operations Manuals.

See Chapter 13 on the Health and Safety File.

5.26 What key duties should the CDM Co-ordinator have completed before a project starts on site?

The majority of the CDM Co-ordinator's duties are undertaken before works commence on site.

The CDM Co-ordinator has to ensure the following:

- All parties who carry out design work on a project collaborate with each other and pay attention to reducing risks to health and safety wherever possible.
- The design is being progressed to avoid foreseeable risks, or where this is not possible, to combat risks at source i.e. that the hierarchy of risk control is followed and/or the principles of prevention followed.
- All Designers co-operate with each other and provide information to each other in order that health and safety risks can be assessed.
- A Pre-Construction Information Pack has been produced and that it has been instrumental in developing the Construction Phase Health and Safety Plan prior to commencement of construction works.
- Advice has been given to the Client if requested, in respect of the suitability of the Construction Phase Health and Safety Plan.
- Advice has been given to the Client on the competency and resources of the designers, and Principal Contractor and other contractors.
- That the F10 has been sent to the HSE, having been updated with additional information, as necessary.

The CDM Co-ordinator should have a major influence over the design team in respect of considering health and safety issues from the design and the construction methods envisaged for the construction phase.

The role of co-ordinating the design team is an important one for the CDM Co-ordinator as it means that all the different designers e.g. architects, interior designers, building services consultants, quantity surveyors and so on should interact with one another to ensure that each of their respective designs do not conflict or create foreseeable health and safety risks.

5.27 A property consultancy acts as managing agent for a large financial institution. Works are to be carried out on one of their buildings and a CDM Co-ordinator has been appointed, but the managing agents are not permitting direct access to the financial institution and so the CDM Co-ordinator cannot communicate with the Client. What can be done?

It will be important first, to determine, who exactly is the Client.

Under CDM 2007, a Client is defined as any person who commissions construction works, either by seeking or accepting the services of another which may be used in carrying out the project for him.

Therefore, the Managing Agent may be the Client if they have instigated works by "seeking or accepting" the services of another to carry out the project.

If the Managing Agent is commissioning the construction works they should be detailed as the Client on the F10.

But, if the building owner has authorized the Managing Agents to approach various contractors to undertake specific construction works which they the owner have specified, then the building owner – in this case the financial institution – will be the Client under CDM and they will need to be detailed on the F10.

Whoever is Client should sign the F10 – or someone else could sign on the Clients' behalf provided they are satisfied that the Client is aware of their duties under CDM 2007.

The CDM Co-ordinator should discuss the CDM Regulations 2007 with the Managing Agents and attempt to ascertain who can legally be defined as the Client. If there is some disagreement, contact the local HSE office and ask for guidance.

The role of the Client is key under CDM 2007 because it is the Client who can have influence over the construction process, health and safety standards, design co-ordination, etc. Duties are placed on Clients because they need to accept responsibilities for health and safety standards on construction projects and they must now ensure adequate time and resources, preparation and planning time, competency of duty holders, availability of information, etc.

If whoever takes on the role of Client fulfils their duties it will not matter whether it is the Managing Agents or financial institution – the intent and purpose of CDM will have been achieved.

But if the Managing Agent does not have the authority to determine preparation and planning time, specifications, budgets and standards then they will not be acting in the role of Client and the CDM Co-ordinator will need to be able to deal directly with the financial institution.

If the Managing Agents fail to facilitate a meeting between Client and CDM Co-ordinator and the Client fails to instruct the Managing Agents to do so then the CDM Co-ordinator is placed in a difficult position.

Where a Client fails to appoint a CDM Co-ordinator they assume that role themselves and must be informed that this is the case.

A project cannot commence without a Client. Someone will have to accept that role and it can only be the person on whom the definition in Regulation 2 CDM 2007 fits.

The best course of action?

- Communication
- Co-operation

Nor resolution? Refuse to accept the commission as CDM Co-ordinator. Professional ethics and reputation has to be worth more than one project fee no matter how tempting it looks.

5.28 Does CDM 2007 apply to term contracts?

CDM 2007 does not apply to the "term contract" but may apply to the individual projects undertaken as part of the term contract.

HSE's general view regarding term contracts is that any F10 notification will be project specific within the general term of the contract.

Notification for a term contract for general work that may not take place is not of benefit.

HSE is interested in works of construction which will last more than 30 days or take more than 500-person days. These projects will be notifiable whenever they occur in a term contract.

CDM 2007 applies to all construction work so really the only consideration will be to decide whether the work package is notifiable.

It may be sensible to address the roles and responsibilities of CDM 2007 and outline the procedural approach for compliance within the term contract.

A practical approach to a term contract in which minor works are carried out to a range of buildings by the same team is to appoint someone into a similar role to that of the CDM Co-ordinator i.e. someone to co-ordinate all the health and safety issues, any design issues across all the contractors teams.

6

Designers

6.1 The CDM Regulations place responsibilities on Designers. Who are Designers?

Under the CDM Regulations Designers are all those who have some input into design issues in respect of a project. These include:

- Architects and Engineers contributing to, or having overall responsibility for, the design
- Building Services Engineers designing details of fixed plant
- Surveyors specifying articles or substances or drawing up specifications for remedial works
- Contractors carrying out design work as part of a design and build project
- Anyone with authority to specify or alter the specification of designs to be used for the structure, including the Client
- Temporary Works Engineers designing formwork and false work
- Interior Designers, Shop fitters and Landscape Architects.

The above includes Architects, Quantity Surveyors, Structural Engineers, Building Services Engineers, Interior Designers, Project Managers (if they can change or issue specifications), Landscape Architects/Designers, temporary Works Engineers designing propping systems, etc.

The "Designer" must be carrying on a trade, business or other undertaking in which he prepares or modifies a design for a structure or arrange or instructs any person under his control to do so.

Anyone who also prepares or modifies a design to a product or mechanical or electrical system intended for a particular structure is also a designer.

Where Clients specify certain sequences of work or the use of specific materials or design features, they will be designers themselves under the Regulations.

6.2 Who decides who is a Designer?

Ultimately only the Court can make definitive interpretation of the Regulations and determine whether an individual or Company is liable for the duties imposed on them.

In reality, you must decide yourself whether you fall within the definition given in Regulation 2 CDM. A designer is any person (including client, contractor or other person referred to in the Regulations) who in the course or furtherance of a business prepares or modifies a design or arranges for or instructs any person under his control to do so in relation to a structure or to a product or mechanical or electrical system intended for a particular structure, and a person is deemed to prepare a design when a design is prepared by a person under his control.

The Client can decide that he expects all his design team to assume "Designer responsibilities" and may conduct competency and resources assessments on all members of the design team. By doing this, the Client will be able to show that he took steps to show that he was "reasonably satisfied" regarding the Designers competency and resources.

It would be prudent to discuss with any professional indemnity insurers their definition of "Designer" under CDM and to ensure that you have appropriate insurance cover for the role you carry out.

6.3 What are the responsibilities of a Designer?

Designers from all disciplines have a contribution to make in avoiding and reducing health and safety risks which are inherent in the construction process and subsequent work e.g. maintenance.

The most important contribution a Designer can make to improve health and safety will often be at the concept and feasibility stage where various options can be considered so as to avoid potential health and safety issues.

Designers must therefore give due regard to health and safety in their design work.

Designers must provide adequate information about health and safety risks of the design to those who need it e.g. proposed roof access routes, use of fragile materials and so on.

Designers must co-operate with the CDM Co-ordinator and other Designers on the project and ensure information is freely available regarding health and safety issues and that they consider the implications of their designs with other aspects of the design, e.g. structural works in relation to building services and so on.

Designers must advise Clients of their duties under CDM, as specified in Regulation 11. CDM requires Designers to take reasonable steps to advise their Clients of the existence of CDM, their duties within the Regulations, the existence of the Approved Code of Practice, good health and safety management and the benefit of making early appointments.

6.4 Are Designers permitted to specify fragile materials?

Fragile materials i.e. materials which give way on impact, point loading and so on are a major cause of both construction accidents and building maintenance accidents. Many of these accidents end as fatalities, or with disabling injuries.

Fragile materials constitute a safety hazard and a design Risk Assessment must be completed which outlines the consideration given in the design process to the use of fragile material, substitute options, etc.

The risks associated with fragile materials are falling through them, injury, death, collapse causing injury to those below, etc.

As the first responsibility of a Designer is to *eliminate* known hazards, the implication is that fragile materials should not be designed or specified into a project.

If the design scheme demands, for example, a glazed atrium roof, then there are inherent hazards with the design. A solid non-fragile material cannot be specified because the glazing is needed to allow for natural lighting. Where such a scenario exists the Designer must specify safety features for the glazed atrium (e.g. safety railings, guard rails, running rail and safety harness fittings, gantry access and so on) and also reduce risk with use of safety glass.

A netting could be designed underneath the glazed atrium to prevent falls should the material give way. Access for cleaning of the glazing, maintenance of the paintwork of the frame, etc. needs to be considered during the design.

If fragile material is used and design considerations allow a *residual* risk, then the Designer must specify information which must be included in the Health and Safety File. The Designer may recommend that a "Permit to Work/Enter" system is adopted by the premises occupier, and that additional safety precautions are needed. The Designer must also consider the safety precautions which the Principal Contractor should adopt so as to ensure safe erection or construction of the fragile material.

6.5 Can the CDM Co-ordinator require changes to designs?

No, unless the Client has included this duty as a specific requirement in their agreement with the CDM Co-ordinator.

However, the CDM Co-ordinator has to ensure that Designers have had adequate regard to their responsibilities for considering health and safety during the project design process, and where they believe the Designer has been reticent they are duty bound to raise their concerns with the Designer and also to advise the Client.

The CDM Co-ordinator should be an expert source of advice about applying the hierarchy of risk control and should be consulted by the Designers whenever necessary.

The CDM Co-ordinator must use best endeavours to advise the Designer if they have concerns about safety issues and must look for alternatives, compromises, etc. If the Designer refuses to listen then the CDM Co-ordinator will need to advise the Client of a potential conflict, particularly if the CDM Co-ordinator believes there will be implications for future occupancy e.g. specifying fragile roof coverings unnecessarily.

If the Designer refuses to listen to the advice offered by the CDM Co-ordinator regarding health and safety in design issues then they may be in a vulnerable position should future accidents occur as a result of the design principles chosen.

A prosecution of the Designer will be possible under Regulations 11 and 18 CDM and the prosecution case would be strengthened if evidence was available that the Designer refused to listen to professional advice.

If the CDM Co-ordinator's advice is ignored, it may provide the CDM Co-ordinator with sufficient evidence to question the competency of the Designer under Regulation 4 and they may advise the Client accordingly, leading to the Designer being dismissed from any approved list and probably, the project.

6.6 What can Designers be prosecuted for under the CDM Regulations?

Any Designer can be prosecuted for not complying with their statutory duties.

If you fail to advise your Client about their duties under CDM you can be prosecuted. If you *do* advise your Client about the requirement to apply CDM to a project and they ignore your advice, provided you have written records that you did everything reasonably practicable to advise and inform your Client of their duties, you may have a defence against prosecution, the HSE preferring to bring the prosecution against the Client.

However, you will need to consider carefully whether you should be working for a Client whom you know to be blatantly ignoring the law. Professional ethics may preclude such an appointment.

If your Client has not appointed a CDM Co-ordinator, nor a Principal Contractor in relation to notifiable projects, nor provides information, then you will have great difficulty in carrying out your own statutory CDM duties.

You can also be prosecuted for failing to comply with your duties to design any construction works safely and with a view to foreseeable risks.

If an accident were to happen on site and the subsequent investigation by the HSE concluded that you, as a Designer, had specified a not commonly known fragile material as a roof covering and that you had not given any information on the hazardous nature of the material to the CDM Co-ordinator or Principal Contractor then it could be argued that the accident was due to your negligence because you had failed to specify safety precautions, or supplied relevant information to allow others to specify safety precautions to be taken.

If, as a Designer, you are insistent on specifying heavy weight blocks or materials for use on the project, and adequate provision has not been made for handling them on site e.g. mechanical aides provided, then you could be liable for any accident or claims.

Equally you could be liable to a prosecution if you specify a substance which is severely harmful, i.e. causes cancer, when there are suitable alternatives on the market.

6.7 Are the CDM Regulations designed to stifle all design creativity?

No. The Regulations impose responsibilities on Designers to consider health and safety issues in respect of their designs.

Many accidents happen on construction sites because Designers give no consideration to the material they specify, the buildability process, the time and resources it takes to do the job. Research by the European Community found that Designers can have a tremendous influence over the number of site accidents and industry ill-health by giving more thought to health and safety in their design.

The CDM Regulations do not intend to bring all building design into grey boring boxes but expects buildings with design flair to be safe to build and for future use.

Designers can still do adventurous things but must consider the practicalities and safety of their designs. Glazed atriums are ideal for some buildings and are perfectly acceptable provided consideration is given to safety issues such as:

- Protection from falling through the fragile material
- Access for cleaning
- Access for maintenance
- Lighting features

Provided solutions are given and the design incorporates these, there is nothing legally that can prevent a Designer designing a glazed atrium.

Buildings can be innovative in design but they need to be practical and safe for all future occupiers. Consider the "Hierarchy of Risk Control" for all design intentions and you will comply with the law.

Case Study

Designers of a new concept retailing outlet sourced a unique mesh type ceiling material which was to be installed in one single sheet to give a ripple effect to the ceiling.

Design considerations included:

- How is it delivered to site
- Will it be difficult to handle
- Will the metal mesh have jagged edges
- How will it be fixed to the ceiling frame
- How will lighting be fixed
- How will sprinklers be fixed
- Will it be a fragile surface
- How could maintenance personnel walk above it to access fittings
- How will it be cleaned.

All of the above, and many more, formed part of the Design Risk Assessment. The CDM Co-ordinator was asked to comment on the information available and to offer any advice.

The new ceiling concept was installed and created the innovative design which the Designers and Client had hoped for. Everyone was happy because it was easily cleaned and maintained and created no specific residual health and safety risks.

6.8 Do duties as a Designer under the CDM Regulations only apply when the project is notifiable?

No. Duties as a Designer under Regulation 11 CDM apply to all projects where there is design. There does not have to be a Client, CDM Co-ordinator or Principal Contractor.

Therefore, for all design commissions you must have a procedure in place for documenting how you comply with the legal responsibilities imposed on Designers as contained in CDM 2007.

6.9 Can the health and safety considerations of designs be left to the CDM Co-ordinator?

No. The legal duty to consider health and safety matters in design rests with *Designers*.

The CDM Co-ordinator can be used for advice on how to apply the hierarchy of risk control and the principles of prevention and may have experiences of what has worked in similar design scenarios.

You must provide information to the CDM Co-ordinator where appointed, when asked to do so. The CDM Co-ordinator has to ensure that Designers co-operate and co-ordinate with each other – sharing information which could have health and safety implications.

The CDM Co-ordinator must ensure that the Designer includes among the design considerations, adequate regard for health and safety issues and therefore has a duty to ask relevant questions to establish that this has been the case. You must therefore expect the CDM Co-ordinator to want to see your Design Risk Assessments or any other records kept by the individual or practice and to discuss them with you, making recommendations if appropriate.

6.10 As Designer, I advise the Client of their responsibilities under the CDM Regulations and yet they refuse to make any of the statutory appointments or address the Regulations. Am I liable to be prosecuted if the project goes ahead?

No. Provided you have fulfilled your responsibility to advise the Client about the CDM Regulations and how, in your opinion, they will be applicable to the project, you will not be liable to a prosecution (unless of course you fail in any other duties imposed on Designers under the Regulations).

It would be advisable to have a system of written notification to a Client that CDM applies to the project. This would demonstrate that you had taken your responsibilities seriously.

A Client who was properly briefed about CDM by the Designer, but who refused to appoint a CDM Co-ordinator or Principal Contractor on a notifiable project, or made the appointment too late, will be in breach of the Regulations and will be liable to prosecution by the HSE.

6.11 What are a Designer's key responsibilities before a construction project commences on site?

The duties of Designers are itemized in Regulations 11 and 18. The predominant duty of a Designer is to manage the process of hazard and risk in respect of health and safety, as they have a responsibility to reduce hazards within their designs and also in respect of materials, methods and processes they specify for the construction phase.

Designers must consider the consequences of their designs in relation to future maintenance and cleaning.

All aspects of hazard and risk in relation to the design must have been considered before works commence on site.

The acceptable way to consider hazards and risks is to produce Design Risk Assessments. These should have been produced before construction commencement and passed to the CDM Co-ordinator or contractor for inclusion in the Pre-Construction Information Pack.

The principles of Regulation 11 must be followed in respect of risk assessments, namely:

- Eliminate the hazards
- Tackle hazards at source by designs which reduce the risks to an acceptable level

- Develop designs which protect all people exposed to the hazard and not rely on control measures which just protect an individual.

6.12 Designers have duties to ensure that their designs meet the requirements of the Workplace (Health, Safety and Welfare) Regulations 1992. What does this mean?

Regulation 11 of CDM 2007 states that where a designer designs a structure for use as a workplace, the designer shall take account of the Workplace (Health, Safety and Welfare) Regulations 1992 which relate to the design of and materials used in the structure.

Designers must take account of risks directly associated with the proposed use of the building or structure, including associated private roadways, pedestrian routes and the risks associated with having to clean and maintain permanent fixtures and fittings.

The Workplace (Health, Safety and Welfare) Regulations 1992 apply to all workplaces and are enforced by local Environmental Health Officers. They require buildings or structures to be safe for their occupiers or visitors e.g. not to have slippery floors, unsafe glazing, unguarded drops, changes in floor level, inadequate temperature facilities, poor ventilation, inadequate traffic routes and so on.

The use of glazing materials could be a major safety issue and whilst the designer may stipulate safety glass so the risk of cuts is minimal, the Workplace Regulations require all exposed glazing where there is a risk of personal injury e.g. people walking in to it because they could not see it clearly, to be suitably covered with manifestations. A surprising number of people walk into glazed panels and breaks their noses and bruise their faces. These types of injury incidents are seen as being avoidable if the Architect/Designer did their job well and complied with the "Workplace" Regulations.

Clients have to be satisfied that any structure designed for use as a workplace has been designed taking account of the provisions of the Workplace (Health, Safety and Welfare) Regulations 1992, which relate to the design of, and materials used in, the structure.

In order to satisfy themselves, the Client may require written confirmation from the Designers that their designs comply with the Regulations.

Such a confirmation can be contained in a pre-start "Permit to Proceed" which must be signed by the designer.

The Client does not have to check the Designer's designs to ensure compliance but must be able to demonstrate that he placed the onus for compliance on the Designer.

6.13 Is there a recognized format for Design Risk Assessments?

No. The CDM Regulations do not specify an exact format which has to be used for Design Risk Assessments. They require merely that hazards and risks in the design are considered and, where appropriate, recorded so that the information can be passed to other members of the design team, the CDM Co-ordinator and all contractors, as appropriate.

Suitable information on hazards and risks can be annotated to drawings if this will give clear guidance to whoever will be reading the drawings. Information can equally be collated in pro forma Design Risk Assessment forms – although the HSE do state that complicated design risk assessments are not expected.

If drawings are designed on Computer Aided Design systems, text can be annotated to drawings to convey hazard and risks and the proposed control measures.

Information on hazards and risks associated with the design of the structure, its construction and subsequent use and maintenance must be freely available to all members of the project team.

No one has a monopoly on the safest way to do things and previous experiences of other team members may add valuable contribution to the debate on health and safety.

Some Clients and CDM Co-ordinators advocate that a project risk register is collated – usually by the CDM Co-ordinator – and that all hazards, risk and solutions are recorded, with dates of actions taken, changes made, etc. Common and expected hazards are not expected to be laboriously recorded – competent contractors should be aware of the dangers of using ladders to work at height. What is more important in the design risk assessment is the thought process the Designer has undertaken to see if works can be undertaken at ground level thus negating the need for ladders and eliminating a site risk. Could staircases be installed early on thereby reducing the use of ladders as a means of access to upper levels? The Designer can influence this process and it forms part of the design risk assessment.

6.14 As a design practice, are we legally responsible for the actions of our employees?

Yes. All employers have vicarious liability for their employees and the duties placed on designers apply to the employer of all employed designers. If designers are self-employed, then they are responsible individually for ensuring that they comply with the law.

Employers must ensure that their employees comply with the laws of CDM.

6.15 As Designers, must we provide information on all hazards and risks associated with the project?

No. The CDM Regulations require designers to take account of significant risks which a competent contractor might *not* be aware of in respect of their designs.

All competent contractors, for instance, should be aware of the hazards and risks of working at heights and the designer does not need to give the Principal Contractor chapter and verse on the safety precautions for working at heights. However, the Designer should consider how to *reduce* the need to work at height and if a particular sequence of construction is envisaged, then the Designer must provide this information to the Principal Contractor, via the CDM Co-ordinator on notifiable projects, and direct to the contractor on other projects.

Providing too much information on all hazards and risks of construction obscures the important detail about significant risks.

Contractors need to know about any specific materials specified or construction sequences planned in order to achieve the desired design effect of the building which they may not be overly familiar with e.g. installation of glazed atria.

6.16 The CDM Co-ordinator insists that we co-operate with other Designers. How can we best achieve this?

Designers have to co-operate with the CDM Co-ordinator and other designers and co-ordinate their work so that the principles of CDM for reducing hazards and risks in construction are met.

The reason that co-operation is needed between designers is to ensure that hazards due to incompatibilities between different design functions are identified and addressed.

Co-operation and co-ordination can be achieved by:

- Regular meetings of the design team and others.
- Agreement on the common principles of hazard and risk assessments and the overriding requirement to reduce risks from design and construction.
- Regular reviews and ongoing design team meetings to address changes in designs, developments in design, etc.
- Co-ordinated design drawings which incorporate all design disciplines e.g. architects, structural engineers, mechanical and electrical services, interior designers.
- Initial site survey meetings and co-ordinated site visits.

Designers need to ensure that their clients are informed of their responsibilities under the CDM Regulations and in particular, that they must ensure that designers have adequate resources to undertake their responsibilities. This means allowing enough time during the project design process for such co-ordination meetings to take place.

6.17 As a Designer, what do I not need to do under the CDM Regulations 2007?

Designers are not required to:

- Take account of risks that were not reasonably foreseeable at the time the design was prepared.
- Provide information about unforeseeable risks.
- Provide information about insignificant risks.
- Specify construction methods unless the design proposed is unusual and a competent contractor may need information e.g. a new construction technique or material used abroad and being introduced into this country.
- Monitor, manage, review or provide any health and safety management function over contractors (other than that required under general health and safety laws or professional ethics).
- Review and report on contractors' health and safety performance.
- Keep copious notes and documents on all aspects of design risk assessment other than conclusions about particular hazards and reasons for their design solutions.

6.18 Building services consultants (or mechanical and electrical design and build contractors) advise that they do not have to produce Design Risk Assessments. Is this true?

Building services consultants, mechanical and electrical design and build contractors, public health engineers, etc. are all classed as designers under the CDM Regulations and therefore the duties imposed under Regulations 11 and 18 apply to them. They must therefore consider health and safety in respect of their designs.

Building services installations are often some of the most hazardous activities on site and it is essential that due consideration is given to the hazards and risks of such processes.

Services designers must provide information to the CDM Co-ordinator regarding their proposed design solutions for the installation of plant and equipment, excavations for services, etc. and the CDM Co-ordinator must ensure that this information is co-ordinated throughout the design team.

The best method for Building Services Designers to provide this information is via Design Risk Assessments.

The true intent of the CDM Regulations can be seen with the co-ordination of building services information into the general design as this will allow any conflicting processes to be considered and resolved prior to construction work commencing e.g. rather than build solid walls and then have to chase out conduits for cabling, the design solution could incorporate hollow walls where the cabling could be run in the void thus preventing the need for chasing concrete and the subsequent health hazards of cement dust, noise and vibration and so on.

Building services Design Risk Assessments are required to be provided to the Principal Contractor so that they can address the identified hazards and risks in the Construction Phase Health and Safety Plan.

Specialist contractors are likely to be appointed to install building services and it is vital that the Principal Contractor has all available information so that he can ensure that hazards and risks which will affect the whole site and all operatives are considered e.g. delivery of large and heavy plant and equipment to site, proposed lifting procedures and so on.

6.19 Do the CDM Regulations 2007 specifically require Design Risk Assessments to be generated on every construction project?

No. The CDM 2007 Regulations actually do not specifically state that Design Risk Assessments have to be produced. The Regulations state that Designers must manage design risks by avoiding foreseeable risks and reducing residual risks so far as is reasonably practicable.

The Designer must take reasonable steps to provide with his designs sufficient information about aspects of the design of the structure or its construction or maintenance (including cleaning) as will adequately assist:

- Clients
- Other Designers
- Contractors

to comply with their duties under the Regulations.

The Approved Code of Practice (L144) states that the information which Designers have to provide to other duty holders can be in any format but it should be brief, clear, precise and project specific.

Information can be notes on drawings, written information provided with the design drawings, formal risk assessment forms and project specific instructions.

HSE Inspectors are not looking for meaningless paperwork – they want to see evidence that Designers have thought about health and safety issues and consequences during the development of the project and that they have done everything that they can to assist in reducing on-site accidents and ill-health and that they have also contributed to a reduction in accidents for maintenance and cleaning workers together with building users.

6.20 Designers cannot start work on a project which is notifiable until the CDM Co-ordinator has been appointed, why?

The CDM Co-ordinator is seen as having a pivotal role in influencing health and safety issues during the design process and in the past, Designers have merrily progressed their designs without anyone raised objective questions about safety. Often, projects would have progressed so far down the design process that even if health and safety improvements to the design were suggested it would be too late to make meaningful changes.

A case of "oh well, we've designed it this way now and it's a bit awkward to change it so we'll stick with it".

The Designer can in fact undertake "initial" design work on a project before the CDM Co-ordinator is appointed but this is fairly limited activity.

The CDM Co-ordinator is to be the Clients "best friend", and one hopes, the Designers, too, and this should allow non-confrontational dialogue about designing out hazards and risks on the project which the CDM Co-ordinator may see but the Designer does not.

By being appointed early on into the design process the CDM Co-ordinator can influence decisions and will be able to ensure that Designers are fulfilling their legal duties, in particular that they are applying the Principles of Prevention to a project.

Principles of Prevention

(a) avoiding risks
(b) evaluating the unavoidable risks
(c) combating the risks at source
(d) adapting the work to the individual
(e) adapting to technical progress
(f) replacing dangerous with non/less dangerous
(g) developing prevention policy
(h) collective/individual protection and
(i) giving appropriate instruction.

6.21 What is initial design work?

Initial design is generally taken to be:

- Appraisal of project needs
- Setting of project objectives
- Feasibility in relation to costs
- Possible constraints on the project
- Possible desktop studies for contaminated land, remedial works
- Likely procurement methods
- Strategic brief
- Confirmation of key project team members (positions rather than names).

The Royal Institute of British Architects would suggest that in their Plan of Work, stages A and B would be initial design work.

Once drawings and specifications are being drawn up and once planning permission is being sought the design has moved into outline design and the CDM Co-ordinator should definitely have been appointed.

The Designer should not progress to these stages unless the CDM Co-ordinator has been appointed.

6.22 Designers best practice tips for managing CDM

(1) Clear policy endorsed at Board level or Senior Partner level on the management of health and safety, including how it will be addressed during the design stages of a project.
(2) Established programme of health and safety training and continual professional development.
(3) A system to demonstrate that design staff have a good understanding of the construction process and a good working knowledge of health and safety legislation and guidance and knowledge of the Workplace (Health, Safety and Welfare) Regulations 1992.
(4) Hazard and risk information in their practice library concerning products and materials regularly specified and used.
(5) Established systems for design reviews, including procedures for communicating with CDM Co-ordinators, other designers, etc.
(6) Lists of key hazards and risks associated with designs which are to be targeted on every project e.g. red, amber, green lists of products/processes to be encouraged/discouraged.

6.23 Design best practice: Things you should do

- Adequate access for construction vehicles to minimize reversing requirements (one-way systems and turning radii)

- Provision of adequate access and headroom for maintenance in plant rooms, and adequate provision for replacing heavy components
- Thoughtful location of mechanical/electrical equipment, light fittings, security devices, etc. to facilitate access and away from crowded areas
- The specification of concrete products with pre-cast fixings to avoid drilling
- Specify half board sizes for plasterboard sheets to make handling easier
- Early installation of permanent means of access, and prefabricated stair-cases with hand rails
- The provision of edge protection at permanent works where there is a foreseeable risk of falls after handover
- Practical and safe methods of window cleaning (e.g. from the inside)
- Appointment of a temporary work co-ordinator (BS5975)
- Off-site timber treatment if hazardous based preservatives are used (Boron or copper salts can be used for cut ends on site).

6.24 Design best practice: Things you should avoid whenever possible

- Internal manholes in circulation areas
- External manholes in heavy used vehicle access zones
- The specification of "lip" details (i.e. trip hazards) at the tops of pre-case concrete staircases
- The specification of shallow steps (i.e. risers) in external paved areas
- The specification of heavy building blocks i.e. those weighing >20 kilograms
- Large and heavy glass panels
- The chasing out of concrete/brick/blockwork walls or floors for the instal-lation of services
- The specification of heavy lintels (the use of slim metal or concrete lintels being preferred)
- The specification of solvent-based paints and thinners, or isocyanates, par-ticularly for use in confined area
- Specification of curtain wall or panel systems without provision for the tying of scaffolds
- Specification of blockwork walls >3.5 metres high and retarded mortar mixes.

6.25 Design best practice: Things you should never do

- Pre-Construction Information Pack not to be issued until detailed struc-tural surveys, asbestos surveys, etc. completed
- Scabbling of concrete ("stop ends", etc.)

- Demolition by hand-held breakers of the top sections of concrete piles (pile cropping techniques are available)
- The specification of fragile rooflights and roofing assemblies
- Processes giving rise to large quantities of dust (dry cutting, blasting, etc.)
- On-site spraying of harmful particulates
- The specification of structural steelwork which is not purposely designed to accommodate safety nets
- Designing roof mounted services requiring access (for maintenance, etc.), without provision for safe access (e.g. barriers).

7

The Principal Contractor

7.1 What is, and what are the duties of the Principal Contractor under the CDM Regulations?

The Principal Contractor is a legal appointment which the Client has to make under the CDM Regulations.

The Principal Contractor must be a Contractor i.e. someone who either undertakes or manages the construction work. A Client who normally co-ordinates construction works carried out on their premises, and who is competent and adequately resourced, can be a Principal Contractor.

The Principal Contractor would normally be a person carrying out or managing the construction work on the project to which they are appointed i.e. the main or managing Contractor. However, where specialist work is involved, it may be appropriate to appoint the specialist Contractor as Principal Contractor as they would be more suited to managing the risks of the specialist activity.

The Principal Contractor has specific duties under the CDM Regulations and ultimately, carries responsibility for site-safety issues.

The main duties can be summed up as follows:

- Develop and implement the Construction Phase Health and Safety Plan.
- Appoint only competent and properly resourced Contractors to the project e.g. specialist Contractors, Sub-Contractors.
- Obtain and check Method Statements from Contractors.
- Ensure the co-operation and co-ordination of Contractors whilst they are on site i.e. control multi-occupied site working.
- Ensure health and safety training is carried out.
- Develop appropriate communication arrangements between Contractors in respect of site health and safety issues.
- Make arrangements for discussing health and safety issues relative to the project.
- Allow only authorized persons onto the site.
- Display F10 on site for all operatives to be able to see the details.
- Monitor health and safety performance on site.
- Pass information to the CDM Co-ordinator Planning Supervisor for the Health and Safety File.

The Principal Contractor is a critical appointment to the construction project as the standard of site safety will be determined by the commitment and competency of the appointment.

7.2 When is the Principal Contractor appointed and how is this done?

The Client has to appoint a Principal Contractor on all notifiable projects and as soon as is practicable after having information about the construction aspects of the project, and in good time, so that the Client can determine the competency and resources of the Principal Contractor.

If the preferred Contractor is known prior to the tendering or negotiating phase of the project, he should be appointed early in the design process.

Usually, the Principal Contractor is appointed as a result of the tendering process and is usually the successful tenderer.

The appointment of Principal Contractor is often assumed, or is a verbal instruction but it is best defined in writing from either the Client, Contract Administrator or CDM Co-ordinator (if requested to do so).

The preliminaries in the Bill of Quantities may stipulate that the successful tenderer will be appointed Principal Contractor. Equally, a statement can be made in the Pre-Construction Information Pack.

The timing of the appointment should allow the Principal Contractor opportunity to develop the Construction Phase Safety Plan. The Client has to allow the Principal Contractor "preparation and planning time" before the commencement of construction works and this must be clearly stated so that adequate time is available for developing the Construction Phase Health and Safety Plan.

Preparation and planning time will vary according to the complexities of the project but the matter should be discussed with the CDM Co-ordinator and any concerns about inadequate provision should be raised with the Client.

7.3 Can anyone be appointed as Principal Contractor?

No. You must appoint only competent persons who normally carry out the business of a Contractor.

A Contractor is defined in Regulation 2 of CDM as:

Any person who in the course or furtherance of a business, carries out or manages construction works.

A Client or any other person could be appointed Principal Contractor if they fall within the above definition.

The traditional "main" Contractor could be appointed Principal Contractor, as could a Design and Build Contractor, Management Contractor, or Construction Project Manager. If a nominated Sub-Contractor or Specialist Contractor has more health and safety experience than the Main Contractor they could be appointed Principal Contractor.

A Principal Contractor does *not* have to be the biggest or Main Contractor on the site, but whoever is Principal Contractor must be able to influence and control site safety throughout the project.

7.4 Can a Client appoint himself as Principal Contractor?

A Principal Contractor must be a "Contractor" i.e. someone who carries on a business in which they:

Carry out or manage construction work.

If the Client has an in-house Project Management Team who will be supervising and managing the construction work, and provided they are competent and have the resources to do so, they could appoint themselves Principal Contractor.

In-house maintenance and facilities management departments could appoint themselves as Principal Contractor if they undertake the works themselves, or if they are managing the construction process.

The complexity of the construction project has to be considered and appointments have to be made only within core competencies e.g. facilities or maintenance teams may not be competent to manage the health and safety of a new-build office complex or an industrial unit constructed with a portal frame.

A Client who appoints the Office Manager as Principal Contractor in respect of a new office extension would be likely to be contravening the Regulations because the Office Manager would not normally carry out or manage construction works – he would not be a Contractor and certainly, is unlikely to be competent.

CDM 2007 places great store on the "competency" of duty holders and research has shown that the more competent i.e. experienced and trained, a person is, the more likely they are to understand hazards and risks and manage them accordingly.

The Site Manager or Agent of the Principal Contractor is key in setting the standards of health and safety for the site and so the Health and Safety Executive (HSE) will expect them to be able to demonstrate experience.

Clients could be prosecuted under Regulation 4 CDM 2007 for appointing incompetent duty holders.

7.5 Do Method Statements/Risk Assessments have to be sent to the CDM Co-ordinator? Is it a legal requirement to do so?

No. There is no duty placed on the CDM Co-ordinator under the CDM Regulations to monitor site safety or to be involved in the development of the Construction Phase Health and Safety Plan.

There is no duty placed on the Client under CDM to monitor site safety or to check whether the Principal Contractor is following the Construction Phase Health and Safety Plan, although the Client has to be satisfied that management arrangements exist throughout the project.

Therefore, there is nothing to be gained by the Principal Contractor sending Method Statements or Risk Assessments to the CDM Co-ordinator *unless* the Client has instructed the CDM Co-ordinator to assist the Principal Contractor in such matters by an agreement outside the duties required by the CDM.

If the Method Statement contains the sequence of construction or similar information, it could be a useful document for inclusion in the Health and Safety File in case of future demolition of the structure, and in this instance, it would be appropriate to send the Method Statement to the CDM Co-ordinator.

7.6 Why should Method Statements/Risk Assessments be completed and when should they be done and for what type of work?

Method Statements are written procedures which outline how a job is to be done so as to ensure the safety of everyone involved with the job, including persons who are in the vicinity.

A Method Statement equates to a "safe system of work" which is required under the Health and Safety at Work Etc Act 1974.

Risk assessments are required under the Management of Health and Safety at Work Regulations 1999, and all employers are required to assess the risks to workers and any others who may be affected by their undertaking.

A risk assessment identifies hazards present, evaluates the risks involved and identifies control measures necessary to eliminate or minimize the risks of injury or ill-health.

A Method Statement can be used as the control measure needed to eliminate or minimize the risks involved in carrying out the job.

As Principal Contractor you should carry out risk assessments for all work activities which you require your *employees* to do i.e. your own employed tradesmen.

Also, as Principal Contractor, you should carry out risk assessments for all those work activities which involve all operatives on site i.e. communal activities e.g. access routes to places of work, delivery of materials, plant and equipment and so on.

As Principal Contractor you should receive risk assessments from all the other Contractors, Sub-Contractors and self-employed tradesmen working on the site. These will tell you what the hazards associated with their tasks are e.g. noise from drilling equipment and will include details of how the risks from the hazards e.g. noise induced hearing loss, can be eliminated or reduced.

When you have reviewed each of the Contractor/Sub-Contractor risk assessments you must consider whether, as Principal Contractor, you need to do anything else to protect other workers in the area, i.e. forbid certain work activities in certain areas and so on. If so, you will need to do an additional Risk Assessment which identifies how as Principal Contractor you are going to manage and control the combined risks of several Contractors.

Risk Assessments need only identify *significant* risks involved in carrying out a work activity. Routine risks and everyday risks such as crossing the road to get to the employee car park need not be included.

Where anything unusual or uncommon is to be undertaken on the site, a risk assessment will be essential. Where works involve significant hazards e.g. working in confined spaces, working at heights, working with harmful substances, then risk assessments are legally required and the control measures identified could be incorporated into a Method Statement which operatives are required to follow.

Case Study

The Principal Contractor was responsible for ensuring the delivery of materials to site. The delivery area incorporated the rear access road which was shared by a neighbouring retail premises. There were hazards to both the site operatives and adjoining tenants from the delivery vehicles and the off loading of materials. Hazards included moving vehicles, restricted access to the roadway for emergency vehicles, off-loading materials from the lorries, dust, noise, falling objects. The risks from the hazards included being knocked over, being hit by materials, noise-induced hearing loss, breathing in dust and exhaust fumes, etc.

The Principal Contractor formulated the risk assessment, identifying the above as the hazards and risks, and determining the control measures needed to eliminate or minimize the risks. These included having a banksman to guide in the delivery vehicles, setting specific delivery times, liaising with the adjoining tenants, providing lifting devices, requiring engines to be switched off during delivery, avoiding reversing vehicles wherever possible, etc. (see Appendix 23A for the Risk Assessments).

The Principal Contractor then prepared a short Method Statement which was given to the Site Foreman to follow when deliveries occurred.

> The preparation of this Risk Assessment and Method Statement
> was the Principal Contractor's responsibility because he had overall
> management control of these activities and could co-ordinate everyone
> else's deliveries to site.

7.7 Can generic Risk Assessments be used as the basis of the Health and Safety Plan?

Yes, but they may not be sufficient for you to demonstrate that you have done everything "reasonably practicable" to ensure the health and safety of all persons at work carrying out construction works.

Generic risk assessments i.e. those which cover the general work activities e.g. bricklaying, form the basis of identifying hazards and risks associated with the job. Provided you develop the generic risk assessment to include any site-specific hazard e.g. carrying out brickwork adjacent to a deep water course, and the additional control measures you intend to adopt, you will have sufficient information to ensure operatives work safely.

When reviewing a Construction Health and Safety Plan the HSE inspector will not be satisfied with a Plan which contains only general information, no matter how thick and impressive the Plan looks. Often, a much thinner and more accurate site-specific Plan will gain praise from the Inspector. They will look for information proportional to the project risks – too much information is confusing and often of little value.

7.8 What other requirements in respect of CDM does the Principal Contractor have to comply with?

In addition to developing the Construction Phase Health and Safety Plan, the Principal Contractor has specific duties laid down in Regulations 22–24 of the CDM Regulations 2007.

These include:

- Liaising with the CDM Co-ordinator.
- Ensuring co-operation between all persons and all Contractors on the site or on adjacent sites where there is an overlap e.g. shared access routes.
- Ensuring that all Contractors and all employees work in connection with the rules contained in the Health and Safety Plan.
- Ensuring that only authorized persons are allowed into the premises where construction works are being carried out.

- Ensuring that the HSE notification of Form F10 is displayed, is in a readable condition and is in a position where anyone involved in the construction works can read it.
- Providing the CDM Co-ordinator with information, particularly if that information would be necessary for inclusion in the Health and Safety File.
- Giving reasonable direction to any Contractor, so far as is necessary, to enable the Principal Contractor to comply with his duties.
- Ensuring that the safety rules contained in the Health and Safety Plan are in writing and brought to the attention of persons who may be affected by them.
- Ensuring that welfare facilities are provided which comply with the requirements of Schedule 2 of CDM 2007.
- Advising contractors the time they will be given for preparation and planning.

In order to be able to ensure co-operation between all Contractors it is necessary to have an understanding of Regulations 11 and 12 Management of Health and Safety at Work Regulations 1999. These Regulations require employers and the self-employed to co-ordinate their activities, co-operate with each other and to share information to help each comply with their statutory duties.

For instance, to be effective, Risk Assessments will need to cover the workplace as a whole and the duty to co-ordinate these will be the Principal Contractor's. Information must be provided by all employers/Contractors so as to enable the Principal Contractor to co-ordinate activities.

Another important aspect of co-ordination and co-operation relates to the use of work equipment and tools which are shared by all Contractors on the site. The Principal Contractor may assume responsibility for provision, maintenance and testing of all common equipment e.g. lifting devices or he may pass the responsibility onto another Contractor. It does not matter *who* does it as long as someone assumes responsibility and everyone else knows who that someone is.

The Principal Contractor will need to request the names of all the people, Contractors, Clients, the Design Team, etc. who wish to visit the site as "authorized persons". The Principal Contractor may authorize them to enter all or part of the site. The Principal Contractor should adopt a formal signing in procedure which all persons should follow. Unauthorized visitors should be accompanied around the site by a trained operative.

Under Regulations 22 and 24 of CDM, as Principal Contractor you have responsibility for ensuring that every Contractor is provided with comprehensive information on the risks to health and safety to all employees and others on the site.

In addition, you must ensure that every employer carries out suitable training for all employees, relative to the works involved, and also, that they provide information relating to health and safety issues.

"Comprehensive" information need not just be in writing – it could be diagrams, drawings, or it could be information in languages other than English.

The most appropriate way to ensure information is available is to include it in the Health and Safety Plan.

Under Regulation 24 of CDM, as Principal Contractor you must ensure that there are procedures in place for any employee or self-employed person to discuss health and safety issues and that there are arrangements for co-ordinating the views of others in respect of health and safety issues.

Depending on the size of the project, all that may be required will be an item for health and safety on a site meeting agenda and a formal process whereby someone can raise health and safety concerns without fear of reprisals.

7.9 What other Legislation in respect of Health and Safety does the Principal Contractor have to comply with?

The CDM Regulations are only one set of Regulations governing safety which are applicable to construction sites.

Other health and safety legislation applicable will include:

- Health & Safety at Work Etc Act 1974
- Health & Safety (First Aid) Regulations 1981
- Control of Asbestos Regulations 2006
- Control of Noise at Work Regulations 2005
- Electricity at Work Regulations 1989
- Management of Health & Safety at Work Regulations 1999
- Provision and Use of Work Equipment Regulations 1998
- Personal Protective Equipment Regulations 1992
- Manual Handling Operations Regulations 1992
- Control of Substances Hazardous to Health Regulations 2002
- Reporting of Injuries, Diseases and Dangerous Occurrences Regulations 1995
- Confined Space Regulations 1997
- The Lifting Operations and Lifting Equipment Regulations 1998
- Control of Lead at Work Regulations 2002
- Regulatory Reform (Fire Safety) Order 2005
- Work at Height Regulations 2005
- Control of Vibration Regulations 2005
- Construction (Head Protection) Regulations 1989.

7.10 What does the Principal Contractor have to provide regarding training for operatives on site?

The CDM Regulations do not actually require the Principal Contractor to provide training (other than to his own employees) to operatives on site but to ensure that every Contractor is provided with comprehensible information on the risks to health and safety from work activities on the site i.e. a suitable site induction.

Comprehensive information is information which is understood by everyone. It is meaningless to issue complex site rules and Risk Assessments if the understanding of written English is poor. Verbal instructions, diagrams, etc. may be more comprehensible.

The Principal Contractor should also ensure that every contractor who is an employer provides any of his employees with information, instruction and training as required by the Management of Health and Safety at Work Regulations 1999, and as required by Regulation 13(4) of CDM 2007.

Site induction training is considered to be the responsibility of the Principal Contractor and the information given should include:

- Site rules
- Emergency procedures
- Fire safety
- Accident procedures
- Permit to Work systems
- Site security
- Welfare facilities on site
- First aid facilities
- Management of health and safety on the site.

All the above information should be included in the Construction Phase Health and Safety Plan and a copy of the Plan should be given to the site foreman of every Contractor on the site.

Whether the Principal Contractor or each individual Contractor carries out site induction training, it is essential that written records of the training are kept and regularly reviewed and updated.

The Principal Contractor will require all Contractors to provide evidence of competency in the various trades and will be entitled to request training certificates for trades such as mobile equipment driving, forklift truck driving, gas fitting works, and so on.

All such documentation should be kept readily available by the Principal Contractor and is best kept appended to the Construction Phase Safety Plan.

The Health and Safety Plan should set out what level of training site operatives are expected to have, who is to have provided it, how often and to what standard.

The Principal Contractor's role in respect of training is a co-ordination role unless he is an employer of his own workforce when the requirements of health and safety training and the provision of information will apply equally to the Principal Contractor as to others.

7.11 Is the construction skills certificate scheme a requirement of CDM 2007?

No, not specifically, although the Construction Skills Certificate Scheme (CSCS) Card does give the Principal Contractor, Client and any other person, evidence that the holder has met certain criteria with regard to health and safety knowledge.

The CSCS Scheme is an independently run and assessed health and safety competency programme which confirms that trades people are competent to undertake their duties and that they have passed a general health and safety test.

The increased requirements for contractor and individual competency under CDM 2007 has encouraged both clients and Principal Contractors to require all contractors and operatives to sign up to the Scheme.

Clients may stipulate that the CSCS Card Scheme shall be adopted for the project and so may Principal Contractors.

CDM 2007 does not specifically require membership of any competency and assessment scheme and individual companies can choose to demonstrate their competency in a number of different ways.

When the HSE inspects a construction site or carries out any investigation into an accident or dangerous occurrence, they will be looking for suitable and sufficient evidence of competency and not necessarily membership of commercial schemes.

Evidence of training, tool box talks, etc. may be just as effective as presenting a CSCS Card.

7.12 The Client has appointed several "Client Direct" appointments who are employed by the Client and required to access the site to undertake works. Are they exempt from complying with CDM?

No. As Principal Contractor you have absolute responsibility for site-safety issues and can specify these in site rules included in the Construction Health and Safety Plan.

Even if the "Client Direct" is not a "Contractor" under the Regulations they will either be employers or employees and as such, have legal duties under the Management of Health & Safety at Work Regulations 1999 to comply with the requirements of the Principal Contractor.

Contractors are defined in Regulation 2 of CDM as:

"Any person who in the course or furtherance of a business carries out or manages construction work."

Construction work includes fitting out, commissioning, alteration, conversion, etc.

Regulation 4 of CDM 2007 states that no person on whom the Regulations place a duty, shall appoint a CDM Co-ordinator, designer, principal contractor or contractor unless he has taken reasonable steps to ensure that the person to be appointed or engaged is competent.

No person shall accept an appointment unless he is competent.

Regulation 4 can be interpreted to include all contractors, irrespective of who has appointed them. So, if they are a contractor, working on a construction site, whether the project is notifiable or not, they must comply with all relevant parts of CDM.

In particular, on notifiable projects, they will come under the control of the Principal Contractor and must be bound by the Construction Phase Health and Safety Plan and Site Rules.

Regulation 10 of CDM applies to all Clients and states that they must not cause or permit any person to work on construction work unless they have been provided with relevant information, including the name of the CDM Co-ordinator and Principal Contractor and the contents of the Health and Safety Plan in any notifiable project. In any event, Regulations 11 and 12 Management of Health & Safety at Work Regulations 1999 apply i.e. co-operation of a multi-occupied site and the appointment of a controlling employer for the site. Information has to be shared regarding Risk Assessments as required under Regulation 10 Management of Health & Safety at Work Regulations 1999.

The Principal Contractor has a duty under Regulation 22 of CDM 2007 to ensure that every contractor at work in connection with the project complies with any rules contained in the Health and Safety Plan. If the Health and Safety Plan contains a rule which states that all Client Direct appointments and any others conducting a trade, business or undertaking, shall provide Method Statements/Risk Assessments to the Principal Contractor prior to commencing work, they must legally comply with the rule.

7.13 What actions can the Principal Contractor take where either Contractors or Client Direct appointments fail to comply with the Site Rules, Health and Safety Plan or CDM requirements

Write to the employer of the Contractors/Client Direct and advise them that they are in breach of their duties under CDM Regulations and other Health and Safety legislation and unless they start complying they will be removed from site, incurring any contractual penalties.

Often non-compliance with a requirement is due to fear or ignorance. Perhaps they do not know how to conduct Risk Assessments. If this is the case, provide information and guidance.

Review your procedures for assessing competency of Contractors. Remember the duty to ensure competency of Contractors rests with *any person*, including Principal Contractors who let subsidiary work packages to Sub-Contractors. Did you know the Contractor could not provide Risk Assessments? If so, why appoint them.

Consider whether you are asking the Contractor to provide more information than is justifiable. Risk Assessments and Method Statements need to be relevant and cover *significant* risks. Requesting meaningless paperwork from Contractors merely compounds their reluctance to produce any.

Issue the contractor with a "yellow" or "red" card i.e. a formal way of letting them know that they have breached site rules.

Ask the CDM Co-ordinator to help encourage the Contractor/Client Direct to comply with their statutory duties. The CDM Co-ordinator could give advice to the Client that the Contractor is not competent and recommend that the Client moves to formally dismiss the Contractor.

Ask the local HSE Inspector for guidance. If the site is complying with CDM then there is nothing to fear in seeking advice from an Inspector on how to improve your health and safety management procedures e.g. improving Contractor compliance.

De-list the Contractor/Client Direct from your approved list.

Whatever steps you take, do not allow the breach of safety management procedures to go unrecorded. Keep detailed records of what actions you took to ensure compliance with site rules, who you spoke to, when, how often, etc. Make sure you have given them the information they could reasonably expect to have regarding the site works e.g. Health and Safety Plan.

7.14 What happens if the Principal Contractor does not fulfil his responsibilities under CDM?

As Principal Contractor, you will be in breach of your duties under Regulations 22–24 and could be prosecuted by the Health and Safety Executive.

Also, contraventions of Regulations 4–6 are offences carrying criminal charges and a Principal Contractor could also be in breach of these if they failed to manage the site effectively.

If you fail to protect unauthorized persons from entering the site you commit a criminal offence. In addition, if that unauthorized person has an accident and decides to sue the Principal Contractor in the Civil Courts, the fact that you have been charged with an offence under Regulation 22 will be "prima facie" evidence of guilt in the civil action.

The HSE will prosecute Principal Contractors who fail to discharge their CDM duties effectively.

Prosecutions are taken either in the Magistrates Court or in Crown Court and fines have the potential to be unlimited.

7.15 What notices under CDM must the Principal Contractor display on site?

The principal Contractor must ensure that a copy of the Notification of Project (Form F10) as sent to the Health and Safety Executive by the CDM Co-ordinator is displayed on the site so that it remains legible and can be read by those working on site.

The Notice can be displayed in the Site Office, but this may restrict the number of operatives who could easily refer to it. An acceptable place to display the F10 would be in the welfare or messing facilities where it would be readily available to all operatives. A copy could be kept in the Site Office or with the Construction Phase Health and Safety Plan.

The location of the F10 Notice needs to be brought to the attention of all contractors working on the site and the easiest way to do this is to inform them during the site induction training.

Principal Contractors may be required to display other signs as required by other health and safety legislation e.g. the Information for Employees poster.

7.16 What type of welfare facilities must be provided by the Principal Contractor?

Regulation 22 and Schedule 2 of the CDM Regulations 2007 set out the requirements for the provision of welfare facilities. The Principal Contractor must comply with the duty to provide suitable and sufficient welfare facilities and this will be determined by a risk assessment. The Client must be satisfied that adequate provision has been made for welfare facilities on the project.

Schedule 2 of the Regulations sets out the principles to be observed and these can be summarized as follows:

Sanitary accommodation

- No numbers are specified but they should be suitable and sufficient for the number of operatives on site.
- Adequate lighting and ventilation must be provided.
- Sanitary accommodation must be kept clean.
- Separate male and female accommodation is not required if each water closet is in a separate room with a door which can be locked from the inside.

Washing facilities

- Washing facilities include wash hand basins and, if necessary due to the type of work, showers.
- Facilities should be in the immediate vicinity of sanitary accommodation and reasonably accessible throughout the site.
- Clean hot and cold, or warm water must be provided, preferably as running water.
- Soap or other hand cleaning chemicals must be provided, together with hand-drying facilities.
- Facilities should be well lit and ventilated, and kept clean.
- Unless washing facilities are provided in individual cubicles, there shall be male and female facilities unless they are used only for washing hands, face and forearms.

Drinking water

- A clean, wholesome supply of water for drinking shall be provided in readily accessible locations.
- Drinking water supplies must be labelled or signed as suitable for consumption.
- Drinking vessels shall be provided if the water is not from a drinking fountain.

Accommodation for clothing

- Facilities need to be made available for everyday clothes not worn on the site and for special clothing worn at work, but not taken home.
- Drying facilities have to be provided for clothing.
- Changing facilities must be provided where specialized clothing is required to be worn.

Rest facilities

- Suitable and sufficient rest facilities shall be provided which include areas for eating and preparing meals.
- Consideration must be given to protecting non-smokers from the harmful effects of tobacco smoke – now superseded by the Smoke Free Legislation.
- Where necessary, facilities for pregnant or nursing mothers must be provided.

7.17 What training must be provided by the Principal Contractor?

The CDM Regulations do not specifically require the Principal Contractor to carry out training for persons other than his own employees.

The Principal Contractor is responsible for ensuring that all contractors working on the site are given comprehensible information about the risks they are likely to face during their work on site.

In addition, the Principal Contractor must ensure that contractors provide their employees with relevant training which covers new or increased risks from working on the project.

Notwithstanding the above, the Principal Contractor may be required to carry out induction training as a requirement of the Client and such a "site rule" would have been included in the Pre-Construction Information Pack and further developed by the Principal Contractor in the Construction Phase Health and Safety Plan.

Induction training arranged by the Principal Contractor may be the simplest way for the Principal Contractor to discharge his duties in respect of making information available to all contractors. An organized, planned induction programme would give the Principal Contractor a mechanism to record attendees and to keep training records. In the event of any accident or incident, the Principal Contractor would have attendance records to show that he had discharged his duties regarding the sharing of information relevant to health and safety risks comprehensively.

7.18 What subjects would be appropriate to include in induction training organized by the Principal Contractor?

The duty of the Principal Contractor is to convey information on risks relating to the carrying out of construction works within the designated site. One of the most effective ways to communicate information is via a structured training programme. Subjects to cover in induction training would be:

- Outline of the project – who's who, Client, Design team, etc.
- Statement on health and safety and commitment of Senior Management to high standards of health and safety
- Site specific risks e.g.:
  - Access routes
  - Contaminated land
  - Overhead power cables
  - Underground services
  - Proximity of water
  - Unstable buildings
  - Hazardous materials e.g. asbestos
- Site specific control measures for identified risks
- Site rules
- Welfare facilities available, location maintenance and cleaning provision, etc.
- First aid facilities
- Accident and near miss reporting procedures
- Emergency procedures e.g.:
  - Fire evacuation
  - Raising the fire alarm
  - Name and address of emergency services
  - Assembly point
  - Fire marshals/wardens
  - Building collapse
  - Flood, chemical escape, gas escapes
  - Release of hazardous substances e.g. asbestos
- Responsible persons
- Requirements for protective equipment and clothing e.g.:
  - Use of hard hats
  - Use of safety footwear
  - Use of ear defenders

- General site-safety controls e.g.:
  - Permit to work
  - Permit to enter
  - Use of banksmen
- Arrangements for communicating with all the workforce in respect of health and safety e.g.:
  - Weekly site-safety meetings
  - Notice board
  - Display of safety notices
  - Use of other aids e.g. visual aids to assist those with language difficulties
- Names of safety representatives, competent persons, etc.
- Site security and access procedures.

All of the above information should be available in the Construction Phase Health and Safety Plan and key details should be displayed on an Information Board at the entrance to the site or in the messing facilities.

Case Study

A new shopping centre was being constructed for a consortium of developers. The developers nominated the main developer as Client under the CDM Regulations 2007 and as Client, the developer appointed a Management Contractor to oversee the entire construction works. The Management Contractor was designated Principal Contractor even though they were not undertaking any actual construction works but because they managed the construction process they met the qualification requirement of CDM for a Principal Contractor.

The Client required the Principal Contractor to have overall responsibility for the site and although each tenant shop fit-out had a main contractor, only the Management Contractor was designated Principal Contractor.

The Client imposed a duty on the Principal Contractor to carry out induction training for *all* persons entering the site.

The Principal Contractor set up a separate training room within his site office compound and via a strict security control point, all persons entering the site for the first time had to report to the training office for induction.

Each induction programme ran for 30 minutes and three sessions were undertaken each day – two in the morning and one early afternoon. If persons wanted to gain access to the site at other times, they had to wait until the next induction session.

Every person going through the induction process was registered and when they had completed the course, they signed a declaration to that effect. These records were kept centrally by the Management Contractor.

Each attendee received a photo ID card which indicated that they had been inducted and when. On the reverse of the ID card were the basic emergency rules of the site.

Each individual main contractor undertaking their own Client's shop fit-out was required to provide additional training to operatives about the specific hazards and risks found on the site.

Each main contractor had to regularly provide information to the Principal Contractor on any health and safety issue in their site which could affect the safety of the whole site e.g. LPG storage. The Principal Contractor then ensured this information was added to its induction training programme.

In addition, the Principal Contractor issued a site directive that every contractor will carry out a tool-box talk every month and issued every contractor with a topic timetable and a 5-minute presentation pack to assist them in delivering a consistent message across the site.

8

Contractors

8.1 What are the main duties of contractors under the CDM Regulations?

Contractors are given specific duties under Regulation 19 of the CDM Regulations as follows:

- Co-operate with the Principal Contractor in order that they both comply with their legal duties under *all* applicable legislation.
- Provide the Principal Contractor with information, including if necessary, risk assessments, on any activity, material, process or task which might affect the health and safety of any person at work carrying on construction works, or of any person who may be affected or which might cause the Principal Contractor to review health and safety across the site.
- Comply with any directions the Principal Contractor gives in relation to site rules or any other health and safety matter which the Principal Contractor has a duty to fulfil.
- Comply with the rules contained in the Health and Safety Plan.
- Provide information promptly to the Principal Contractor in relation to accidents, diseases or dangerous occurrences, including any fatalities, which the Contractor would need to report under the Reporting of Injuries, Diseases and Dangerous Occurrences Regulations 1985.
- Advise the Principal Contractor of any incident or of any situation in which it will not be possible for him to comply with the Construction Phase Health and Safety Plan or Site Rules.
- Provide information to the Principal Contractor on any matter which it might be reasonable to assume the Principal Contractor will need to pass onto the CDM Co-ordinator for the Health and Safety File.

In addition to the above, no contractor shall start any construction works unless they have been provided with the nature of the CDM Co-ordinator and the Principal Contractor i.e. have been given, or have seen, a copy of Form F10.

All contractors must have relevant parts of the Construction Phase Health and Safety Plan as relates to the work they will carry out on the site.

In addition to the specific duties relating to notifiable projects, contractors are "persons" under the Regulations and therefore have a duty to ensure that they provide information to others, only appoint competent persons to carry out tasks, co-operate and co-ordinate works with all others to whom it may be relevant.

In certain circumstances, contractors may also be designers and will need to comply with the specific duties placed on designers.

8.2 What responsibilities do contractors have for on-site training of their operatives when the project is non-notifiable?

Where contractors are employers they have general duties under health and safety legislation to ensure that their employees receive suitable and sufficient training, or information, instruction and training.

A general duty for training employees is contained in Regulation 13 of the Management of Health and Safety at Work Regulations 1999 especially where they are likely to face new or increased risks within the work environment.

Regulation 13 of CDM contains specific duties for contractors in relation to providing information and training to every worker carrying out construction work under his control, so that the worker can carry out the particular work safely.

Contractors will need to provide any worker on any project which is not notifiable with:

- Induction training
- Information on the risks to their health and safety
- Information on their risk assessments provided under Regulation 3 Management of Health and Safety at Work Regulations 1999
- Information relating to the activities of any other contractor working on the site which may be relevant to ensuring his health and safety
- Any site rules
- Emergency procedures, including fire
- Names of the people who are responsible for managing the site or carrying out certain tasks.

The above will probably best be covered by a pre-start meeting and formal training session. Simple site rules can be issued and various posters displayed in the mess room, office, etc. to advise workers of the safety controls in place on site.

As the contractor has a duty to ensure that every worker is informed of relevant site safety matters, it would be sensible to ensure that a record of attendance is kept for every induction or meeting. Workers should preferably sign their attendance.

8.3 The Principal Contractor has not organized induction training. What should contractors do?

Good health and safety starts with communication and co-operation so the first step to take is to discuss the matter with the Principal Contractor.

Review the Construction Phase Health and Safety Plan to see how the process of induction training was to be implemented and inform the Principal Contractor if there seems to be a discrepancy.

Under Regulation 24 CDM, the Principal Contractor must ensure that there is a mechanism for employees and the self-employed at work on the construction site to discuss and offer advice in respect of health and safety.

Failure to have an appropriate induction training programme will affect the health and safety of the site and its operatives. The Principal Contractor has to *ensure* that contractors have provided their employees with information about risks, control measures and emergency procedures.

If the Principal Contractor still fails to fulfil his legal duties, the next step is to seek advice from the CDM Co-ordinator regarding the competency of the Principal Contractor. The CDM Co-ordinator has a duty to provide advice to any contractor with a view to them complying with their duties in competency and resources, and in turn, if information is made available to the CDM Co-ordinator which questions the competency of the Principal Contractor, the CDM Co-ordinator will have a duty to inform the Client.

Contractors are entitled to receive parts of the Construction Phase Health and Safety Plan which directly affect them and this would include information on welfare facilities, fire precautions, emergency planning, accident management, site security, etc. If this information is made available to the contractors, they should be able to organize their own induction training rather than leave their operatives without it.

In some circumstances, the Principal Contractor may provide the relevant information to the contractor's site foreman for them to give suitable "tool box talks" on-site induction information.

Contractors must remember that the duty to ensure that their employees are adequately trained rests with them as employers and that this duty cannot be delegated to someone else.

CDM 2007 increases the expectation on all persons, and especially, duty holders to co-operate, co-ordinate and communicate with one another.

The CDM Co-ordinator's role is specifically to help co-ordinate the flow of information across the project, including any information required for the construction phase.

Also, the Client, and the CDM Co-ordinator, need to be satisfied that management arrangements are in place for the duration of the project and it should become very evident to them early on in the project if the Principal Contractor is failing in their duties.

8.4 Do contractors have to provide risk assessments to the Principal Contractor?

The Principal Contractor has a duty to co-ordinate health and safety across the construction site and, in particular, must address hazards and risks which affect all operatives on the site, no matter who their employer is.

Under Regulation 19 a contractor must provide relevant information to the Principal Contractor on any activity which might affect the safety of operatives or others working on or resorting to the site. Relevant information includes any part of a risk assessment made under the general provisions of the Management of Health & Safety at Work Regulations 1999.

Principal Contractors should not really require risk assessments for tasks which only affect the contractor's operatives but often, in order to ensure that a culture of health and safety pervades the site, the Principal Contractor may include a Site Rule which states that risk assessments for all activities must be provided to them. Contractors have a duty to comply with site rules.

8.5 What are some of the practical safety initiatives which I can introduce to the Construction project in my role as Principal Contractor?

Site safety does not stop with compliance with the CDM Regulations 2007. Legislation really only sets minimum standards to be achieved and good safety management endeavours to achieve higher standards. Good safety management has been proved to have positive benefits to employers, contractors, Clients and employees and ultimately reduces costs due to accidents, incidents, down time, stoppages, investigations, loss and damage to property, equipment, etc.

In order to encourage high standards of safety on construction sites, the Health and Safety Executive supports a campaign known as "Working Well Together".

This campaign encourages all those involved in a construction project to work together to reduce hazards and risks and improve overall, standards of health and safety in construction.

The following pages outline some of the initiatives which have been, or can be, introduced during a construction project.

8.5.1 Preventing falls from heights

The Principal Contractor undertook the following actions:

- Influenced designers to schedule permanent access structures e.g. staircases early in the project.
- Provided communal mobile tower scaffolds for use by all contractors and assumed responsibility for daily checking and remedial repairs by competent persons.
- Provided full edge protection to all drops more than 2 metres.
- Included within the Site Rules procedures for safe working at heights.
- Introduced a Permit to Work System for all work to be undertaken above 2 metres.

- Required all contractors to provide risk assessments and a work plan for all activities they proposed to undertake above 2 metres.
- Required all contractors to undertake tool box talks on working at heights, the hazards, risks and controls necessary and required evidence from each contractor of such training or instruction.
- Provided a site audit function to check regularly on activity.

8.5.2 Preventing slips, trips and falls

The Principal Contractor introduced the following:

- Information on slips, trips and falls, types of injury sustained, etc., included with site induction training.
- Clearer emphasis within Site Safety Rules on good housekeeping, trailing cables, etc.
- Provision of adequate numbers of transformers, etc. in location near to where power tools are to be used.
- Regular site safety inspections.
- Employed a site labourer, dedicated to clearing the site of debris, etc.
- Adequate provision for skips, etc. on site for containing waste materials, etc.
- Regular maintenance of floor surfaces, infilling of holes, etc.
- Specific tool box talks with electricians so that they did not leave floor box recesses unguarded.
- Provision of good, overall site lighting and where necessary, appropriate task lighting.
- Assessment of operative's footwear.

8.5.3 Improvements in manual handling

The Principal Contractor introduced the following:

- Mechanical lifting devices available to all contractors on the site, co-ordinated by the CDM Co-ordinator, maintained and certified by him as necessary.
- Overall assessment of what materials were needed to be delivered to site, where they needed to be used, where they could be stored, etc. This entailed requesting from all contractors plans of work and scheduled deliveries of materials, etc.
- Inclusion within site induction training of a section on manual handling, hazards and risks and aspects of manual handling that have been identified within the site, using a training video.
- Requirement for all contractors to provide risk assessments and method statements for manual handling activities.
- Alterations to specifications of materials so that smaller and less heavy sizes were delivered to site.
- Poster campaign around the site.

8.5.4 Painting and decorating

The Principal Contractor was aware that the majority of painters and decorators on site were small companies or self-employed individuals. Their general awareness of health and safety was poor and other trades were complaining to the site agent. The Principal Contractor decided to implement an awareness campaign on health and safety for the painters and decorators and introduced the following:

- Tool box talks by the Principal Contractor's Site Safety Officer on:
  - Working at heights
  - Working with hazardous substances
  - Working with power tools
  - Working in close proximity to other trades and people
  - Housekeeping
- Requirement for method statements and risk assessments for all work activity.
- Easy access to mobile tower scaffolds provided by the Principal Contractor.
- Leaflet campaign on contact dermatitis of hands and arms.
- Early morning work co-ordination meetings to agree who would be working where, what the hazards would be, etc.

8.5.5 Improving health and safety awareness of site foremen and supervisors

The Principal Contractor included a section in the Construction Phase Health and Safety to improve health and safety awareness of all contractor and subcontractor site foremen and supervisors. During the tendering process, each contractor had to agree to sign up and commit to the site foreman initiative.

The Principal Contractor then introduced the following:

- Special training sessions for all site foremen on the roles and responsibilities for managing health and safety.
- Regular site safety inspections carried out with each site foreman.
- A safety review meeting held weekly and attended by all site foremen and supervisors.
- Accident prevention campaign which highlighted a safety topic every month and reviewed incidents and accidents.
- Guidance and best practice booklets produced by the Principal Contractor's own Safety Department for distribution to each site foreman.
- Site-specific safety audit checklists which each site foreman was required to complete and return to the Principal Contractor.
- Introduction of a "Safety Default Notice" system which recorded poor standards of health and safety against each contractor.

8.5.6 Site-safety campaign

The Client on a major retail store development project was the instigator of a site safety campaign which was operated by the Principal Contractor on the

Client's behalf. The best performing contractor in respect of health and safety was awarded a monetary sum for donation to a charity of their choice.

The Principal Contractor decided that the award should be given to the contractor who had made the most and greatest contribution to overall site safety. This involved not only ensuring that their own operatives were highly safety conscious, but they also conducted their undertaking in such a way that the safety of others was improved.

The Principal Contractor arranged information meetings with all the contractors on site and set out clearly the performance criteria expected.

Assessments would be made on the following:

- Frequency and efficiency of tool box talks.
- Practical, working knowledge of operatives on site on general health and safety principles.
- Quality and effectiveness of method statements and risk assessments.
- Co-operation and co-ordination of work with other contractors.
- General commitment to health and safety, demonstrated by pro-active attendance at the weekly site-safety meeting.

The Principal Contractor provided the services of the on-site Safety Officer to any contractor who wished to have extra assistance in upgrading their health and safety practices. The Principal Contractor provided all contractors with a basic training pack for tool box talks and attended many as an observer. Contractors were required to provide evidence of tool box talks and individual operatives were interviewed to establish their learning outcomes.

The campaign ran for a 3 month period and the Principal Contractor and the Client, assisted by the CDM Co-ordinator, assessed all information, accident records, etc. and decided on an outright winner.

A presentation award ceremony was arranged in the site canteen and the Client's Director presented a cheque for £500 to the successful contractor. In turn, they presented a cheque to a children's disability charity.

Similarly, on a major pharmaceutical company's research and development campus construction site, the Managing Contractor, as Principal Contractor, implemented an "Accident-Free Working Hours" campaign and set an objective to reach 1 million working hours free of any major accident or incident. The Site Safety Officer ran numerous campaigns on the site on safety topics, supported by poster campaigns, training, site audits, feedback to contractors, etc. When milestones of accident-free working hours were met, all, operatives on the site were awarded some tokens of achievement such as t-shirts, mugs, kit bags and so on. When a major milestone, such as 500,000 accident-free working hours, was achieved, a major "reward" was issued, such as fleece jackets.

Although the campaign in itself might have been an additional cost on the project, the benefit was immense, because the entire project lost no down time for accident investigations, stoppages, poor operative performance, etc. The project came in on time and predominantly on its half billion pound budget!

8.6 Contractors have to check that Clients are aware of their duties. What does this mean?

Contractors are expected to be competent individuals or organizations with knowledge of relevant legislation, good practice and industry guidance as appropriate for their trade or business.

Clients may be very inexperienced in dealing with construction projects and may not be aware of any specific legal duties placed on them when commissioning construction work.

Contractors are therefore expected to be able to check with a client that they know what the legal requirements are, and in particular, that they know what their duties are under CDM 2007.

Contractors should ensure that Clients know about appointing competent contractors or others, obtaining and sharing information, appointing a CDM Co-ordinator and Principal Contractor if the project is notifiable, etc.

Contractors should not start work on a project unless they are satisfied that their Client is competent and understands his duties under CDM 2007.

A simple, standard checklist to issue to the Client would be an effective procedure to put in place and the letter/checklist could be sent to all Clients at the time of tenders, quotes, specifications, etc.

Then, if a Client were to say that they were unaware of their duties and they had not been informed, you would have a record to the contrary.

9

Design Risk Assessments

9.1 Designers have duties under CDM 2007 to avoid foreseeable risks to the health and safety of people constructing their design, affected by the construction works or using or resorting to the building or structure once built. What does this mean?

Designers must consider the health and safety aspects of their designs and must avoid foreseeable risks i.e. those risks which their experience, knowledge and competence tells them that they should know about.

Many accidents occur in buildings because the end design is not user friendly, or the materials looked good when first built but after some wear and tear they become hazardous, or the access to plant and equipment is poor and maintenance personnel have to undertake hazardous tasks to do their jobs, etc.

Designers have great influence on the future safe use of the building and CDM 2007 requires them to recognize this fact by considering health and safety.

Designers can also have great influence on the safety of contractors by paying attention to health and safety when they choose materials, building location, site set out, access, construction sequences, etc.

Designers can do more to ensure that contractors and their operatives work more safely on construction sites by ensuring that they specify more appropriate materials, equipment, etc. e.g. pay greater attention to size, weight.

If the risk is foreseeable i.e. you could see or know about it beforehand, then CDM 2007 requires the Designer to avoid it.

9.2 Risks may be foreseeable but they may not always be avoidable. What happens then?

Risks will always be with us and it will be impossible to eliminate all risks. Health and safety legislation is qualified by the term "as far as is reasonably practicable" and this means that a judgement can take place.

If a risk cannot be eliminated completely, it may be able to be replaced by something with a lesser risk e.g. a substance known to cause cancer could be replaced with one known to cause skin irritation. It would still be a risk but the consequences of use are less and perhaps more acceptable.

Health and safety law refers to:

- Hierarchy of risk control
- Principles of prevention.

Designers would have to demonstrate that whenever they are unable to eliminate a hazard and risk they have ensured that they have chosen a design with the minimum risk or that they have protected against the hazard and risk.

9.3 As a Designer, do I have any responsibility for site safety matters during construction?

No. There is no legal responsibility on Designers to undertake any safety auditing role.

Obviously, frequent visits to site will give ample opportunity to identify any breaches of health and safety rules and where these are noted they should be reported to the Site Agent for their action.

Professional conduct codes require members to act responsibly when safety breaches are encountered and to report them. You do not have to tell the Site Agent what to do to put them right unless you have specific information available about a material or product.

Health and safety legislation imposes a duty on all persons to take responsibility for reporting or taking action in respect of health and safety hazards or contraventions.

9.4 As the Designer, what are my key responsibilities under the CDM Regulations during the construction phase of the project?

The predominant duty of a Designer under CDM is to ensure that due regard is given to the health and safety impact of their designs.

Designers need to follow the hierarchy of risk control, namely:

- Eliminate the hazard and risk at source
- Substitute for a lesser risk
- Protect the whole workforce
- Protect the individual.

Often, the design process carries on during the construction phase with Designers making alterations to designs and site conditions may vary to what was expected. In these instances, the Designer must ensure that revisions to drawings or schemes consider health and safety implications, where appropriate, new design risk assessments will be required. These should be passed to the CDM Co-ordinator so that they can be integrated into the Construction Plan, which the Principal Contractor has a duty to ensure is reviewed, amended and updated during the course of the project.

The Designer must provide the CDM Co-ordinator with information on the design, specifications, etc. for the Health and Safety File. Revisions to schemes should therefore be recorded so that accurate information is available for future reference.

Designers should take responsibility for ensuring that design information, etc. is made available to the CDM Co-ordinator during the course of the project – do not wait until completion. Materials, datasheets, revised drawings, operating and maintenance manuals can all be passed on when available. This will enable the CDM Co-ordinator to review the information and request further details if necessary so as to enable them to produce a comprehensive and meaningful Health and Safety File for the project.

9.5 Does my liability as a Designer under CDM finish when the construction project has ended?

No. The CDM Regulations require you as a Designer to consider the future use of the building, including maintenance and cleaning and future demolition.

You must therefore apply the hierarchy of risk control to future maintenance and cleaning issues e.g. access to plant and equipment, maintenance activities, cleaning of high level windows, glass atria, high level shelving, light fittings and so on.

If an accident were to happen to a maintenance worker, for instance they fell from height, because as a Designer you had not provided a safe means of access to their place of work e.g. no guarded gantry walkway, accessible ladder and so on then an investigation by either the Health and Safety Executive (HSE) or Local Authority Environmental Health Officer (EHO), depending on the type of premises, could conclude that the accident happened because of poor design principles which did not consider health and safety.

As Designer, you could be interviewed to determine what "design risk assessments" you carried out on your design and be asked to explain how you arrived at the design decision you made.

No legal precedent has yet been set in respect of a Designer prosecution for accidents caused once the building is occupied but as both the HSE Inspector and the EHO become more familiar with applying the CDM Regulations to accident investigations the more likely a test case will be taken.

Designers already have civil liabilities in respect of their designs being "fit for the purpose" – CDM will impose criminal liabilities.

9.6 Who is responsible for ensuring that Design Risk Assessments are provided to the CDM Co-ordinator?

Designers, Principal Contractors and contractors all have duties to ensure that the CDM Co-ordinator is provided with information for the Health and Safety File.

Once the building or structure has been built, the Designer will be responsible for ensuring that the CDM Co-ordinator is provided with information regarding any residual health and safety risks, in particular, in relation to future maintenance, cleaning and demolition.

Information can be provided in the form of Design Risk Assessments or it can be provided in the form of annotated drawings, detailed specifications, manufacturer's operations and maintenance manuals, a risk register.

The Principal Contractor must provide design information to the CDM Co-ordinator in respect of any design detail which either the Principal Contractor or other contractors have completed on site i.e. ongoing design details.

If for example, a specialized curtain walling contractor has installed glazed walls, then there will be design considerations and specifications regarding its stability, erection details, maintenance and cleaning and so on. The specialist contractor would provide this design information to the Principal Contractor who would pass it on to the CDM Co-ordinator.

The CDM Co-ordinator may visit site during the construction phase, in their co-ordination role, and seek out relevant information rather than waiting for it to be provided.

Designers pass information to the CDM Co-ordinator on design issues, Principal Contractor passes information on construction issues and contractors pass information usually to the Principal Contractor for onward transmission to the CDM Co-ordinator.

CDM Co-ordinators not only need Design Risk Assessments but also *any* information which they deem may be appropriate for inclusion in the Health and Safety File.

9.7 What consideration does a Designer need to give to "cleaning issues"?

Designers have to consider the health and safety implications of "cleaning work" within the structure once it is handed over to the Client.

"Cleaning work" in respect of design considerations involves the cleaning of any window, translucent or transparent wall, ceiling or roof, in or on a structure.

Information on the design considerations of cleaning parts of the structure must form part of the Designer's Risk Assessment and this must be included in the information for the Health and Safety File.

Many accidents occur to employees and maintenance workers once a building is occupied because the Designer "has forgotten" to design in a safe system of work for reaching surfaces which need to be cleaned.

Designers must give thought to the process and need to design in mechanisms for safe systems of work e.g.:

- Permanent edge protection
- Overhead gantries
- Vertical cradle systems
- Fall arrest systems
- Access routes for "cherry pickers"
- Frequency of cleaning.

If mechanical systems are unnecessary, then the Designer needs to ensure that they have allowed storage space for stepladders or ladders.

The practical solutions intended to be used by the Designer need to be included in the Health and Safety File. This information should then be accessed by the building occupier's employees or maintenance contractors, thereby enabling them to undertake the cleaning work safely.

Case Study

A major refurbishment was carried out to a licensed premises which had originally been built 5 years previously.

The original building complied with Building Regulations and CDM 1994. The roof was three storeys up and flat with no edge protection. The original Designers had installed a fall arrest system, operated via a "running line" around the roof. Over the years statutory inspections by the Local Authority had approved the safe system of work on the roof.

The new Designers specified for new plant to be put on the flat roof and in so doing they reviewed the fall arrest system already installed. They considered it suitable, had it checked and re-certified and included the information in the Health and Safety File. Maintenance teams, etc. were re-trained and a Roof Access Permit was implemented.

But, during the final stages of the Project the HSE Inspector arrived on site for a routine CDM visit. Site safety was generally good and there were no issues – until the Inspector went on the roof.

He was less than satisfied with what he saw and was unwilling to accept the fall arrest system as a suitable design solution. He wanted to know why the Designers – the Project Architects – had not replaced the running line with a full edge protection systems.

The Inspector formed the view that the Designer had not considered the "Principles of Prevention" and had not eliminated a hazard which was foreseeable and reasonably practicable to do.

The Designer was left trying to justify his decisions to leave the fall arrest system in place – but he had no documented records of his review procedures nor evidence to justify his design decision.

9.8 What is initial design work?

Initial design is generally taken to be:

- Appraisal of project needs
- Setting of project objectives
- Feasibility in relation to costs
- Possible constraints on the project
- Possible desktop studies for contaminated land, remedial works
- Likely procurement methods
- Strategic brief
- Confirmation of key project team members (positions rather than names).

The Royal Institute of British Architects would suggest that in their Plan of Work, stages A and B would be initial design work.

Once drawings and specifications are being drawn up and once planning permission is being sought the design has moved into outline design and the CDM Co-ordinator should definitely have been appointed.

The Designer should not progress to these stages unless the CDM Co-ordinator has been appointed.

10

The Pre-Construction Information Pack

10.1 What is meant by pre-construction information?

Any information which could be important in assisting designers or contractors in carrying out their duties under CDM 2007 could be classed as pre-construction information.

The information required under CDM 2007 is really only that which is rele-vant to health and safety so that hazards and risks can be identified and subsequently addressed.

The Client is responsible for deciding what information is available and what can be provided to Designers and Contractors.

Where projects are notifiable, the CDM Co-ordinator can advise the Client has to what information will be needed and assist the Client in obtaining the information.

10.2 Who has to provide the pre-construction information?

The Client has the duty to provide information to every person designing the structure and to every contractor who has been or may be appointed by the Client.

When a project is notifiable, the Client has to provide the information to the CDM Co-ordinator and the CDM Co-ordinator has to pass it on to every person designing the structure and to any contractor, including the Principal Contractor who has been or may be appointed by the Client to the project.

Designers have to provide information to the Client, other designers, the CDM Co-ordinator and other contractors as is necessary for them to carry out their duties. Some of this information will need to be included in the Pre-Construction Information Pack.

10.3 Is there a specific format for the Pre-Construction Information Pack?

No. The CDM Regulations do not specify the format that the pre-construction information should be in and indeed, does not really refer to a "pack" as such.

Information could be available in a wide variety of locations and there is no requirement to duplicate the information. HSE are keen to see the amount of unnecessary paperwork on CDM projects reduce significantly.

Whatever information is available and wherever that information is located, it is important that it is clearly identified and its availability made clear to those who may need it.

An information "road map" could be created, listing what is available, where and who holds it.

Where projects are notifiable, the CDM Co-ordinator could advise the Client early in the project on the format and content of the Pre-Construction Information "Pack".

Information packs should not be generic with lots of "strike throughs" and irrelevant information.

10.4 What type of information specifically is the Client required to provide?

The pre-construction information shall consist of all the information in the Client's possession, or which is reasonably obtainable, including:

- Any information about or affecting the site or the construction work
- Any information concerning the proposed use of the structure as a workplace
- Any information about the existing building.

The Client must ensure that he has obtained anything about the site, including environmental issues, local traffic, use of the site, previous use of the site, use of adjoining premises/land, overhead power cables, public transport and so on.

Designers and contractors will need to know what the proposed use of the building is, especially if it will be a workplace because designers, particularly, will need to ensure that they design the structure in accordance with the Workplace (Health, Safety and Welfare) Regulations 1992.

10.5 Who has to be provided with the pre-construction information?

Very broadly, every person designing the structure and every contractor who has been or may be appointed by the Client.

Designers will be:

- Architects
- Quantity surveyors
- Structural engineers
- Surveyors.

- Civil engineers
- Interior designers
- Landscape architects
- Specialist contractors e.g. temporary works
- Design and build contractors.

Contractors will be:

- Any of those who are tendering
- Specialist contractors
- Demolition contractors
- Mechanical contractors
- Electrical contractors
- Ground work contractors
- Civil and structural contractors.

The list of contractors who may be appointed could be considerable and the requirement for information could generate vast amounts of paperwork.

The Client, and on notifiable projects, the CDM Co-ordinator, could identify key information which contractors will need to adequately quote for the work required.

Once a short list of tenderers has been drawn up, or one contractor chosen, the Client could discuss further details of the information which would be specific to the execution of the works.

10.6 Does a Client have to provide any additional information to a contractor?

Yes. A new requirement of CDM 2007 is that the Client must advise the contractor – or any contractor – the minimum amount of time before the construction phase which will be allowed to the contractors appointed by the Client for planning and preparation for construction work.

Contractors will need to determine the resources they need to carry out the project safely and they therefore need to have an idea of the time available for further research and mobilization for the project.

Preparation and planning is not defined in the Regulations but is taken to mean the time period in which the contractor considers exactly how to do the job, exactly what equipment will be needed, what specialist contractors, etc. need to be procured, what welfare facilities are required and how they will be installed on the site, the number and skill base of operatives, etc.

Unrealistic deadlines and failure to allocate sufficient resources – both financial and human – are two of the main causes of construction site accidents and poor risk management.

Clients should consult with the CDM Co-ordinator on notifiable projects and also with all contractors and discuss the time scales needed, expected and offered.

Naturally, clients will often believe that contractors are prolonging the period they need for mobilization and for the construction phase, and they will be keen to ensure a shorter timescale as possible.

Contractors often say that clients, or indeed, their advisors do not understand the full complexities of what they have to do to mobilize and complete a project.

If a contractor says the project can be completed in 24 weeks, the Client will often counter that it has to be finished in 20 weeks, believing that the contractor will be "dragging his heels for 4 weeks".

Sensible dialogue is necessary and a true consensus of opinion and expectation needs to be reached.

Again, the CDM Co-ordinator could assist in these discussions, and if the project is not notifiable, and the Client is unsure of what timescales would be reasonable, the Client could consult the competent person appointed under the Management of Health and Safety at Work Regulations 1999.

10.7 How thorough does the Client or CDM Co-ordinator have to be in providing information?

The answer from the HSE is likely to be "very".

The provision of comprehensive and suitable information about potential and actual hazards and risks relating to health and safety on the project is seen to be a key weapon in ensuring better standards of health and safety.

The Client and CDM Co-ordinator will be expected to do everything which is reasonably practicable to obtain the information which could protect the health and safety of those constructing the building/structure or working in or on it.

It will no longer be acceptable for Clients to leave the contractors to find things out for themselves – they must make all reasonable enquiries to obtain the information.

If information is not available, the Client must commission surveys to obtain the information.

As an example, many buildings may contain asbestos containing materials and these are known to be a serious health hazard. Often, asbestos is hidden in buildings and its presence not easily identifiable.

Exposure to asbestos is a major health hazard and when contractors do not know whether they will be exposed to the substance, they cannot plan the job safely and may release fibres affecting themselves and others.

Clients are expected to provide thorough and detailed information on the likelihood of asbestos containing materials and should commission surveys if they do not have any information available in an asbestos register.

Surveys need to be thorough and whenever any demolition works (of any sort) are planned a Type III, destructive survey, must be commissioned.

If a Client is seen to only commission a presumptive, visual Type I survey, this will not be seen as suitable and sufficient information and the Client may well be in breach of the duties imposed on them in Regulations 10 and 15 of CDM 2007.

Clients, or the CDM Co-ordinator, should search archives, records, ask maintenance managers, other contractors, etc. whether they hold records of drawings, specifications, etc.

Utilities companies should be contacted for details of underground and overground services.

Contaminated land surveys, radon surveys, atmospheric surveys, etc. may be needed.

The Client would be wise to communicate with his Design Team and the CDM Co-ordinator (for notifiable projects) about the type of information they believe would be necessary and to "brainstorm" what other information might be available.

Information will also be required about the minimum standards the Client will expect the contractor to meet in relation to managing health and safety. Such standards are often known as Employers Requirements and include standard operating procedures for permit to work systems, accident reporting, fire safety, induction training, etc.

10.8 Will non-notifiable projects require less detail in the Pre-Construction Information Pack than notifiable projects?

No, not necessarily. Regulation 10 of CDM 2007 applies to all projects and requires that the Client provides every person who may design the structure and every contractor who has been or may be appointed to the project to be provided with all the information in the Client's possession (or which is reasonably obtainable), including:

- Any information about or affecting the site or the construction work
- Any information concerning the proposed use of the structure as a workplace
- The minimum amount of time before the construction phase which will be allowed for preparation and planning
- Any information in any existing health and safety file.

The CDM 2007 Regulations apply to all construction projects and so whatever information is relevant to the project must be made available or commissioned.

Notifiable projects require the Client to provide the information to the CDM Co-ordinator and cite the same level of information as required for non-notifiable projects.

The HSE do expect the information provided to be proportionate to the risks involved in the project and so will not expect the level of detail in the information pack to be the same in all projects.

But small non-notifiable projects for instance, carried out in buildings which continue to operate as businesses may have greater hazards and risks than a green field new build site. The expectation will be that detailed information will be provided to the contractor on the refurbishment project, including on-site health and safety procedures, traffic management, permit to work systems, fire safety, deliveries, access for personnel, site security, residual site hazards and so on.

Case Study

A CDM Co-ordinator acts for one Client on many similar projects.
A Pre-Construction Information Pack format was agreed as a "core" document for all projects, with one section dedicated to the site-specific information.

The contents of the Pre-Construction Information Pack included:

- Introduction and purpose of the plan
- Project details and information
- Timescales for preparation and planning
- Health and Safety objectives for the project
- Responsibilities of all parties
- Statutory requirements applicable to the project
- Project specific health and safety information
- Site rules and management procedures
- Requirements for managing health and safety resources, contractors, etc.
- Requirements for site welfare facilities
- Requirements for co-ordination and co-operation regarding design changes, design risks
- Project review procedures
- Information for the Health and Safety File.

10.9 What type of information should be available from the Client?

The Client has a duty to provide information to all Duty Holders in order to enable them to fulfil their legal CDM duties.

Information will vary according to the complexity of the construction project e.g. a green field site may have little available information whereas a refurbishment of an existing building will require detailed knowledge of the structure, surrounding area, building use and so on.

The Client should be able to provide the following information, arranging to have surveys carried out if necessary:

- Contaminated land surveys
- Existing services locations
- Structural/building safety reports
- Survey reports for hazardous substances e.g. asbestos, lead, toxic substances
- Survey reports for hazardous areas e.g. confined spaces
- Survey reports for hazardous locations e.g. fragile roof access
- Local environmental conditions
- Local hazardous areas e.g. schools, major roadways
- Current Health and Safety File
- Intended occupancy details
- Proposed site rules e.g. existing Permit to Work systems
- Proximity of water courses, transport systems, etc.
- History of any previous damage e.g. fire damage, floods and so on.

The Client should be encouraged to commission surveys when the information produced would be vital to the planning of the project in respect of health and safety. The HSE will expect Clients to have taken responsibility for identifying hazardous conditions or substances e.g. asbestos and would consider prosecuting where there has been a blatant disregard to making information available.

Case Study

The Principal Contractor was undertaking refurbishment works when he came across unexpected asbestos lagging to pipe work, previously hidden by a partition wall.

The project had to stop whilst notification for asbestos removal was made to the HSE The HSE Inspector visited site and wanted to know why the required 14 Day Notice had not been submitted. He formed the view that no planning for asbestos had been undertaken and that the Principal Contractor was trying to "pull a fast one" to avoid the 14 day delay before works of removal could commence.

The HSE Inspector inspected the Pre-Construction Information Pack and noted that an asbestos survey had been carried out but that the asbestos lagging on hidden pipes had not been noted and was unforeseen. The Principal Contractor had the current methods in place for the removal of any other asbestos found in the building and the Inspector was reasonably satisfied.

However, the investigation turned to the CDM Co-ordinator who was asked to account for the actions the Client took regarding commissioning asbestos surveys and how the Client judged the competency of the asbestos surveying Contractor. The Inspector formed the view that it was reasonable for the Client to have commissioned a more detailed survey which should have included for the removal of wall and ceiling panels to ascertain whether asbestos was in any hidden areas, particularly in view of the buildings age.

The Inspector issued an advisory letter to the Client stipulating that reasonable enquiries should have been made to obtain relevant information and that in his opinion, the Client had failed to comply with the duties imposed on him in Regulations 10 and 15 of CDM 2007. The letter went on to say that any further breaches of the Client's duty to provide information relating to a construction project would be dealt with by more formal action.

10.10 Does reference have to be made in the Pre-Construction Information Pack to "good practice" and legal requirements?

The Pre-Construction Information Pack does not need to repeat legislation applicable to construction safety but it can include references to good practice where the Client and CDM Co-ordinator (if appointed) expect such standards to be adopted into the Construction Phase Health and Safety Plan.

It may be helpful to include references to legislation where unusual safety hazards exist or where legislation has recently been introduced with which the Contractors may not be fully conversant e.g. Work at Height Regulations 2005.

The Client or CDM Co-ordinator may have decided to include references to HSE Guidance Notes or Codes of Practice within the information pack or to other trade or professional guides. Of particular importance may be standards on Fire Safety and reference to the Loss Prevention Council's code of practice on fire safety may stipulate the minimum standard of fire safety on the site which the Contractor should price for.

Rather than include detailed references in the body of the Pre-Construction Information Pack it may be more practical and beneficial to include an Appendix to the document which lists all relevant Legislation, Codes of Practice and Guidance available to the project.

Experience has shown that the Contractors are sometimes unaware of the extent of health and safety legislation and appreciate advice and information. The CDM Co-ordinator can create and develop a good working relationship with the Principal Contractor if they are perceived as helpful and supportive, and likewise the Client with Designers and contractors on non-notifiable projects.

10.11 What does the Pre-Construction Information Pack need to include in order to comply with the Regulations?

The Pre-Construction Information Pack should include the following:

- General description of the construction works comprised in the project.
- Details of the time within which it is intended that the project, and any intermediate stages, will be completed.
- Details of the Project team, including CDM Co-ordinator, Designers and other Consultants.
- Details of existing plans, any Health and Safety File, etc.
- Details of risks to health and safety of any person carrying out the construction work so far as such risks are known or are reasonably foreseeable or any such information as has been provided by the Designers, including Design Risk Assessments.
- Details of any Client requirements in relation to health and safety e.g. safety goals for the project, site rules, permits to work, emergency procedures, management requirements and so on.
- Such information as the Client or CDM Co-ordinator knows or could ascertain by making reasonable enquiries regarding environmental considerations, on-site residual hazards, hazardous buildings, overlap with the Client's business operation, site restrictions, etc.
- Such information as the Client or CDM Co-ordinator knows or could ascertain by making reasonable enquiries and which it would be reasonable for any Contractor to know in order to understand how he can comply with any requirements placed on him in respect of welfare by or under the relevant statutory provisions e.g. availability of services, number of facilities required to be provided, details of any shared occupancy of the site.
- Content and format of the expected Health and Safety File.

The Approved Code of Practice gives guidance as to what to include in the document and this can be summarized as:

- Nature and description of project
- Client considerations and management requirements
- Existing environment and residual on-site risks
- Existing drawings
- Significant design principles and residual hazards and risks
- Significant construction hazards.

The Information Pack is expected to include only that information which it is reasonable to expect or which could reasonably be ascertained by making enquiries. The Regulations do not expect every aspect of potential hazard and risk to be known at the outset of the project but expects reasonable enquiries to be made.

The *competency* of the Duty Holders in foreseeing hazards and risks of the site, the existing building or the design and future use of the building would be investigated if obvious health and safety risks were not highlighted in the Pre-Construction Information Pack.

10.12 It has become apparent that further investigative works are required on the site to establish whether the land is contaminated. As Client, am I responsible for arranging this?

The question to really ask here is why the likelihood of a contaminated site was not discovered prior to the commencement of the construction phase. The Client is responsible under CDM for providing all persons with information regarding the state or condition of any premises at or on which construction work is to be carried out. Where the project is notifiable, the CDM Co-ordinator has a duty to relay this information to the Principal Contractor in the Pre-Construction Information Pack.

Should evidence be discovered that potentially contaminated land exists, the Principal Contractor should take steps to ensure the health and safety of all operatives and persons resorting to the site.

The Client will be responsible for commissioning a suitable survey so that information on hazards and risks can be provided to the Principal Contractor. The Client may prefer to commission the Architect, Project Manager or Principal Contractor to organize the investigation works, but whether they do so or nor, the Client is responsible for ensuring that they do not abdicate their responsibilities for health and safety within the project process and must ensure that the Principal Contractor has adequate resources and the competency to deal with the hazards discovered.

If the project is notifiable, the CDM Co-ordinator should give advice to the Client on the steps to take to ensure that any unexpected hazard is properly managed and that information is communicated to the people who need to know.

Should the information come to the attention of the HSE inspector, they will be particularly interested to establish what actions the Client took prior to the commencement of works. They would also be interested in the role of the Design Team and the CDM Co-ordinator and will be looking to see what information was available at design stage and what additional surveys would have been reasonable to have carried out.

11

The Construction Phase Health and Safety Plan

11.1 What is the Construction Phase Health and Safety Plan?

The Construction Phase Health and Safety Plan is the document produced by the Principal Contractor which develops any information contained in the Pre Construction Information Pack and sets out the arrangements the Principle Contractor will take to ensure management of health and safety on the site.

The document is required by Regulation 23 CDM 2007.

The Health and Safety Plan is the foundation upon which the health and safety management of the construction phase needs to be based. A written plan clarifies who does what, who is responsible for what, what hazards and risks have been identified, how works shall be controlled, etc.

The Contractor appointed Principal Contractor must develop the Construction Phase Plan *before* construction works start so that it outlines the health and safety procedures which will be adopted during the construction phase.

The Construction Phase Health and Safety Plan must be site specific and reflect the site hazards and risks. It needs to be focussed and suitable for the works to be undertaken. Generic, weighty documents written by people who have never been to the site are not recommended and Health and Safety Executives (HSE) are increasingly identifying these Plans as being unsuitable, leading them to serve Improvement Notices under the Health and Safety at Work Etc Act 1974.

Appendix 3 CDM 2007 lists the information which should be included in Construction Phase Health and Safety Plan as follows:

- Description of the project
- Details of the project team
- Details of any existing information and location of records, etc.
- Management of the work including roles and responsibilities of key people
- Goals and objectives regarding health and safety for the project
- Health and safety arrangements for the project once on site:
  - Liaison and co-operation between all parties
  - Consultation procedures with the workforce

  - Exchange of design information
  - Handling design changes during the progress of work
  - Selection and control of contractors
  - Exchange of health and safety information
  - Site security, including preventing unauthorized access
  - Site induction
  - Site training
  - Welfare facilities
  - First aid
  - Accident reporting
  - Procedures for preparing, issuing, reviewing risk assessments and method statements
  - Site rules
  - Emergency procedures
  - Fire safety arrangements
- Arrangements for controlling significant site risks:
  - Delivery procedures
  - Removing waste
  - Location, management of services
  - Stability of structures
  - Contaminated land issues
  - Environmental conditions
  - Overhead, underground power supplies
  - Temporary structures
  - Work at height
  - Work on or near fragile materials
  - Hazardous substances
  - Control of lifting operations
  - Maintenance of equipment
  - Excavations, earthworks, tunnels, etc.
  - Demolitions
  - Work near water courses
  - Use of compressed air
  - Working with explosives
  - Traffic routes and vehicle management
  - Storage of materials
- Arrangements for controlling health hazards on site:
  - Asbestos
  - Hazardous materials
  - Manual handling
  - Noise
  - Stress
  - Exposure to radiation
- Details to be included in the Health and Safety File, including layout, format of information, timescales to provide information, etc.

11.2 What format does the Construction Phase Health and Safety Plan have to have?

The Construction Phase Health and Safety Plan should be in a format which is:

- Easy to use and to refer to
- Understandable to those who need to use it
- Easy to update
- Easy to duplicate
- Clear, concise and logical.

The Construction Phase Health and Safety Plan is needed *on site* and should therefore be robust and almost non-destructible. An A4 ring binder is a popular choice for keeping the information in order, as pages can be easily removed and photocopied for other Contractors as necessary and importantly, it can be easily updated.

The Construction Phase Health and Safety Plan would not be particularly useful on computer disk on the site because access to its information may be restricted.

The Construction Phase Health and Safety Plan need not be all written words – often the use of diagrams, pictograms and cartoons are very effective at explaining health and safety messages.

The Plan should not contain every health and safety procedure ever produced but only those *applicable* to the site. This should prevent it from becoming unwieldy and will not discourage people from reading it. It is perfectly acceptable to The Health and Safety manual, or to the COSHH (Control of Substances Hazardous to Health Regulations 2002) manual. If site-safety procedures are reliant on safety procedures specified in other documents then copies of these must be available on site.

The Construction Phase Health and Safety Plan should not contain all the site Risk Assessments and Method Statements, nor quantities of generic forms.

The Construction Phase Health and Safety Plan should be a "roadmap" to point all operatives on site to the key information on how health and safety will be managed on the site.

11.3 Does a copy of the Health and Safety Plan have to be given to every person working on the site, the Client, the CDM Co-ordinator, etc.?

The Client must be given a copy of the Plan so that they can be satisfied that it has been prepared and complies with the Regulations.

There is no duty to give a copy to the CDM Co-ordinator unless they are acting on behalf of the Client in assessing its adequacy before construction works can start. In this case, there would be no need to send one to the Client, unless specifically required to do so.

The CDM Regulations require the Principal Contractor to provide information to all persons working on or resorting to the site in respect of health and safety issues.

As the Health and Safety Plan contains valuable information on-site health and safety matters it makes sense to issue the document to as many people as practicable. However, that may become expensive and some operatives may only be on site for a few days. Information could be issued to the Site Foreman. Relevant information could be displayed around the site in poster format.

Site-safety rules should be issued to all individual operatives and should be issued to all employers/Contractors/employees during site induction training.

Specific Risk Assessments should be issued to the ganger or foreman, with instructions that he is responsible for ensuring that all his gang/team are made aware of hazards and risks and the protective measures needed to control the risks.

Copies of relevant information could be displayed in the mess room, site office and site canteen e.g. location of first aid kit, names of companies with trained first aiders.

Key aspects of the Health and Safety Plan can be issued to safety representatives, site foremen, etc., with guidance on how and where they can access the full Health and Safety Plan and supporting information, documentation e.g. Company Safety Policy, HSE Codes of Practice and so on.

A practical way of disseminating site health and safety information is to convene a weekly site-safety forum or committee, requiring a foreman or representative from every Contractor or self employed person on site to attend, using the meeting to review site health and safety issues and to discuss forthcoming works on the programme, new site-safety rules, etc.

It is important to remember that CDM 2007 places a duty on all persons to ensure co-operation, co-ordination and communication between all persons engaged in the construction project.

There will be no one set way to disseminate the information in the Construction Phase Health and Safety Plan and it will be for the Client, the CDM Co-ordinator and the Principal Contractor to discuss the plan and the methods of sharing the information during any preparation and planning time.

The Construction Phase Plan must be site specific i.e. it must cover issues which apply to the works to be carried out, include actual site personnel, site-specific emergency procedures and so on.

HSE are increasingly identifying Construction Phase Plans which are generic and serving Improvement Notices on the basis that the plans are not suitable and sufficient.

Appendix 3 of the Approved Code of Practice for the CDM Regulations lists the information which should be included in the Construction Phase Plan as follows:

- Description of the project.
- Arrangements for the project (including, where necessary, for management of construction work and monitoring compliance with the relevant

statutory provisions) which will ensure, so far as is reasonably practicable, the health and safety of all persons at work carrying out the construction work and all persons who may be affected by the work of such persons at work, taking account of:

(i)   The risks involved in the construction work.

(ii)   Any activity of persons at work which is carried out, or will be carried out on or in premises where construction work is undertaken.

(iii)   Any activity which may affect the health and safety of persons at work or other persons in the vicinity.

- Sufficient information about arrangements for the welfare of persons at work by virtue of the project to enable any Contractor to understand how he can comply with any requirements placed upon him in respect of welfare by or under the relevant statutory provision.

A Construction Phase Health and Safety Plan outline is shown in the Appendices at 11.A.

Case Study

A major shop fitting Contractor acted as Principal Contractor and developed a comprehensive Health and Safety Plan. A summary version was produced which covered Site Rules, Emergency Procedures, Fire Safety, Personal Protective Equipment, Signing in Procedures, etc. and this document was given to all operatives and visitors to site who underwent the site-safety induction training. Additional notices were displayed in the canteen, mess room, site office and at the entrance to the site.

Each major Contractor on site was given a full copy of the Health and Safety Plan and was required to sign a record to that effect. They were then required to ensure that all relevant information regarding health and safety had been provided to their site operatives.

The Principal Contractor introduced an auditing system which regularly checked how, to whom and when the Contractor issued the information.

The Principal Contractor provided a number of site posters in a variety of languages and with cartoons, pictograms, etc. and encouraged all contractors to take note of them. This helped disseminate key health and safety information to all operatives and met the Principal Contractors' duty to ensure that information was suitable for those who had difficulty in understanding English.

11.4 As Principal Contractor, I intend to revise the original Construction Phase Health and Safety Plan. Do I need to tell the CDM Co-ordinator or the Client?

Only if your revisions are due to design changes. The CDM Co-ordinator has to ensure that Designers conduct Risk Assessments and need to be sure that these have been done. They rely on the Principal Contractor to inform them. It should be beneficial to have a discussion with the CDM Co-ordinator regarding the proposed changes.

There is no duty to advise the Client of any changes to the Construction Phase Health and Safety Plan under the CDM Regulations and there is no duty on the Client to check that you have made changes or updates to the Plan, although they must be satisfied that you have procedures in place to revise, amend and monitor the Plan.

However, you will want to create a good impression and demonstrate competence so it would be advisable to keep the Client informed. This could be done via Project Meetings and it would be wise to minute a Health and Safety section on the agenda.

The Client and the Design Team or other professional advisers will need to be briefed on any changes to the Construction Phase Health and Safety Plan which could affect their safety e.g. changes to site rules, restricted areas of the site and so on.

11.5 What is the best way to update the Construction Phase Health and Safety Plan without it becoming complicated or confusing?

The Construction Phase Health and Safety Plan is the document which sets out the health and safety management of the construction phase of the project. It must be a document which outlines what special health and safety precautions are to be taken to ensure the safety of everyone on the construction project and must be updated to reflect any changes to the working procedures, management systems, welfare facilities, etc., which happen on the site.

If the majority of the Construction Health and Safety Plan has been agreed before the commencement of the construction works there will be little need to change substantial parts of it. Details on site management, emergency procedures and welfare facilities may not change during the construction phase if they have been well thought through at the beginning.

If the Construction Health and Safety Plan needs to be updated it should be done by adding information clearly and removing old information so as to avoid confusion. For instance, the names of the trained first aiders may change. The new ones should be added to the Plan and the old ones removed. If the location of the first aid kit has changed this should be included.

The most important thing about the Construction Health and Safety Plan is that the information contained in it is made available to all operatives on site – the simpler the updates the easier things will be understood.

Risk Assessments and Method Statements could be included as a separate document to the Plan, making it easy to add new information to Risk Assessments but without changing the overall Plan.

If the Principal Contractor decides to implement a new Permit to Work procedure for a specific activity which has only recently come to light, then this Permit to Work system must be clearly explained in the Construction Health and Safety Plan.

An aspect of the Construction Health and Safety Plan which will need to be constantly kept under review, and updated when necessary, will be the Fire Safety Plan. As construction work progresses, site exit routes may become altered e.g. by permanent partitioning and so on. Alterations must be clearly depicted on the Fire Safety Plan.

The Principal Contractor should ensure that, perhaps once a week, time is set aside to review the Construction Health and Safety Plan and any relevant changes which are made must be *communicated* to site operatives via the arrangements made for ensuring health and safety issues are considered e.g. at the weekly site contractors' meeting.

Feedback from site operatives on health and safety matters should be considered and the Construction Health and Safety Plan amended or updated to take into account operatives concerns, ideas and suggestions as to how the site could be improved from a health and safety point of view.

Generic Risk Assessments will need to be reviewed and updated to incorporate site-specific issues. These should then be kept in a separate document to the Construction Health and Safety Plan, together with any associated Method Statements. Individual Risk Assessments can be issued to specific operatives as necessary, or importantly, to the Contractor Foreman so that he can assess what safety precautions need to be followed by his team.

11.6 Is the Fire Safety Plan a separate document?

No, the Fire Safety Plan can be an integral part of the Construction Phase Health and Safety Plan.

The Fire Safety Plan should identify fire risks throughout the site e.g.:

- Combustible materials
- Use of hot flame equipment
- Use of liquid petroleum gas
- Use of combustible substances

- Storage and use of any explosive materials and substances
- Sources of ignition e.g. smoking
- Use of heaters.

Once the potential fire risks are identified i.e. where, when, why and how a fire *could* start on site (or the surrounding area, yards, outbuildings) the Fire Safety Plan should include precautions and procedures to be adopted to *reduce* the risks of fire. These could include:

- Operating a Hot Works Permit system
- Banning smoking on site in all areas other than the approved mess room
- Controlling and authorising the use of combustible materials and substances
- Providing non-combustible storage boxes for chemicals
- Minimising the use of liquid petroleum gas and designating external storage areas
- Controlling the siting and use of heaters and drying equipment
- Operating a Permit to Work system for gas and electrical works.

Having identified the potential risks and the ways to minimize them, there will always be some residual risk of fire. The Fire Safety Plan should then contain the Emergency Procedures for dealing with an outbreak of fire, namely:

- Types and location of Fire Notices
- The location, number and type of fire extinguishers provided throughout the site
- The means of raising the alarm
- Identification of fire exit routes from the site and surrounding areas
- Access routes for emergency services
- Procedure for raising the alarm
- Assembly point/muster point.

The Fire Safety Plan should also contain the procedures to be taken on site to protect against arson e.g.:

- Erection of high fencing/hoarding to prevent unauthorized entry
- Fenced or caged storage areas for all materials, particularly those combustible
- Site lighting e.g. infra-red, PIA.
- Use of CCTV
- Continuous fire checks of the site, particularly at night if site security is used.

Procedures for the storage and disposal of waste need to be included as waste is one of the highest sources of fire on construction sites.

Materials used for the construction of temporary buildings should be fire protected or non-combustible whenever possible e.g. 30 minute fire protection. The siting of temporary buildings must be considered early in the site planning stage as it is best to site them at least 10 metres away from the building being constructed or renovated.

Having completed the Fire Safety Plan, a sketch plan of the building indicating fire points, assembly point, fire exit routes, emergency services access route to site, etc. should be completed and attached to the Plan. The sketch plan (which could be an Architect's outline existing drawing) should be displayed at all fire points and main fire exit routes and must be included in any site rules/information handed out at induction training.

11.7 What criticisms does the HSE have of Construction Phase Health and Safety Plans?

The HSE have raised many concerns about the quality of Construction Phase Health and Safety Plans, in particular regarding the general content which is often not relevant to the project in hand. They would prefer thinner but more site-specific documents.

Some of the common deficiencies are itemized as follows:

- Activities not assessed i.e. those activities with health and safety risks which affect the whole site or specific trades e.g. storage and distribution of materials, movement of vehicles, pedestrian access ways, removal of waste, provision and use of common mechanical plant, provision and use of temporary services, commissioning and testing procedures, etc.
- Management arrangements do not focus sufficiently on the role of Risk Assessments.
- Site supervisors and managers do not have reasonable knowledge of safety, health and welfare requirements and standards.
- Site supervisors and managers are not familiar with the contents of the Construction Phase Health and Safety Plan.
- Monitoring arrangements are overlooked or the "competent" person performing this role is not suitably qualified.
- Details of welfare provision are limited to a couple of lines of the Plan. It should cover in explicit detail the requirements and implementation of Schedule 2 Construction (Design and Management) Regulations 2007.
- Fire precautions, including arrangements for the fire alarm system (if required) and emergency lighting are often overlooked.
- The implication for health and safety of tight timescales for the project are not fully addressed in the Plan. The Plan often fails to recognize that shortening a construction programme increases the amount of material stored on site, increases operatives on site, all of which leads to restricted work space, inadequate supervision, poor co-ordination and control, etc. This issue should now be clearly set out in the section of the Plan which covers "preparation and planning time".

The HSE Inspectors believe that all of the above must be considered before works commence on site and that if a Construction Phase Health and Safety Plan does not adequately address them, a Client should not approve the document under Regulation 16 of the CDM Regulations.

11.8 What other Health and Safety legislation must the Construction Phase Health and Safety Plan address?

The Construction Health and Safety Plan is a specific requirement of the CDM Regulations but the CDM Regulations are not the only health and safety legislation applicable to construction projects.

The Construction Phase Health and Safety Plan must make reference to the monitoring of compliance with the relevant statutory provisions which are applicable to the construction site, namely but not exhaustively, the following:

- Health & Safety at Work Etc Act 1974
- Health & Safety (First Aid) Regulations 1981
- Control of Asbestos Regulations 2006
- Control of Noise at Work Regulations 2005
- Electricity at Work Regulations 1989
- Construction (Head Protection) Regulations 1989
- Manual Handling Operations Regulations 1992
- Personal Protective Equipment Regulations 1992
- Gas Safety (Installation and Use) Regulations 1998
- Reporting of Injuries, Diseases and Dangerous Occurrences Regulations 1995
- Health & Safety (Consultation with Employees) Regulations 1996
- Health & Safety (Safety Signs – Signals) Regulations 1996
- Control of Lead at Work Regulations 2002
- Confined Spaces Regulations 1997
- Provision and Use of Work Equipment Regulations 1998
- The Lifting Operations & Lifting Equipment Regulations 1998
- Management of Health & S afety at Work Regulations 1999
- Control of Substances Hazardous to Health Regulations 2002
- Work at height Regulations 2005
- Regulatory Reform (Fire Safety) Order 2005.

Not all of the above legislation will apply to a construction site – it depends on the complexity of the project and the type of work to be carried out. The Principal Contractor must be aware of which Regulations are applicable to the works and must ensure that both their own employees and other Contractors are complying with the requirements where necessary.

If the Principal Contractor is not familiar with the requirements of the legislation he should seek expert guidance from in-house safety officers or external Consultants. The CDM Co-ordinator should be able to give good practical guidance if asked to do so.

The duty to comply with some of the Regulations may fall to the Principal Contractor where the activity will affect all operatives and visitors to site e.g. conducting a COSHH Assessment on dust which affects the whole of the site.

Case Study

The Principal Contractor identified that several trades would be using power tools at the same time and although individual Contractors had provided ear defenders to their own employees, there were other operatives in the vicinity who would be subjected to high noise levels over a prolonged period. The Principal Contractor carried out a noise assessment of all the tools in operation at any one time and concluded that the noise level in the site was over the legal limit, and that he must therefore take action to reduce to noise levels. He re-organized the work programme so that only half the number of tools were in use at any one time, thereby reducing the overall noise level to below the legal limit. This action was preferable to issuing all workers on the site with ear defenders i.e. he had controlled the noise at source – step one of the Hierarchy of Risk Control.

12

Competency of Duty Holders

12.1 What is meant by the term "competency"?

Competence means the ability to perform any activity required to the level of performance expected.

To be competent, an organization or individual must have:

- A good understanding of the hazards and risks associated with the tasks they carry out
- Sufficient knowledge of the tasks to be undertaken
- Sufficient expertise and ability to carry out their duties in relation to the project
- An ability to recognize their limitations and when the job demands more skill than they have
- An ability to know what control measures need to be put in place to manage the risks of the tasks being carried out.

12.2 Is competency a "once and for all" standard?

No. The development of competence is a life long learning process. Individuals continue to develop competency as they become more experienced, more knowledgeable and better trained.

The type of equipment in use is construction projects, the type of materials which can be specified, work practices and procedures, etc. all constantly evolve and operators must be brought up-to-date with new techniques, etc.

Competency may have been evaluated many years ago when equipment or tools were basic but with the use of modern methods of construction, etc. the operative may no longer be competent.

The saying "but we've always done it this way" does not necessarily mean that competency can be taken for granted.

12.3 What are the requirements of CDM 2007 regarding competency?

Regulation 4 CDM 2007 sets out very clearly, and very early on in the Regulations, that people appointed to various duty holder positions must ensure that the people they appoint are competent to carry out the duties given to them.

The Regulation covers all persons who may have duties under the Regulations i.e. "no person shall appoint ....".

The Regulation specifically concerns itself with the appointment of:

- CDM Co-ordinator
- Designer
- Principal Contractor
- Contractor

i.e. the four duty holders identified in the Regulations.

Not only must the person who makes an appointment make sure that the appointee is competent but the person accepting the appointment must only do so if they are competent.

No person shall arrange or instruct a worker to carry out a design or manage a design, or undertake to manage construction work, unless the workers is:

- Competent or
- Under the supervision of a competent person.

12.4 Is competency restricted just to health and safety matters?

CDM 2007 is particularly targeted at ensuring that health and safety standards in the construction industry improve and so competency is focussed on assessing health and safety knowledge and experience.

But if individuals do not understand and have experience in the trade or profession they are practising then they are unlikely to be able to demonstrate competency in health and safety.

If, for instance, the Client wishes to appoint a Structural Engineer they will need to ensure professional competence before assessing health and safety competence – a Structural Engineer who did not understand load bearing structures would be incompetent even if he had a knowledge of health and safety.

12.5 How should a Client assess the competency of the CDM Co-ordinator?

The role of the CDM Co-ordinator is seen as pivotal to good health and safety management on a construction project and if the post is to have real benefits it needs to be carried out by an organization or individual who really understand the importance of the role.

CDM Co-ordinators have to encourage co-operation between individuals, designers, contractors and others and so will need good inter-personal skills.

Good co-operation is essential and the CDM Co-ordinator has a duty to ensure that it happens amongst all duty holders on a project.

CDM Co-ordinators should have:

- Good communication skills
- Common sense
- An ability to listen
- An ability to problem solve
- An ability to mediate
- An ability to get to the hub of the issues and not to be side tracked by "red herrings".

CDM Co-ordinators should not rely on over bureaucratic procedures to protect their own position and really must foster good information flow across all elements of the project.

So, the first stage for the Client is to establish whether they could work well with the CDM Co-ordinator. Would they trust their judgement, professional opinion, advice? If they can establish that rapport easily then the chances are everyone else in the project team will as well.

The next stage is for the Client to assess the CDM Co-ordinators knowledge of the design process.

The CDM Co-ordinator will need to be able to work closely with the design team and will need to understand the design process, the inter-relationship between disciplines, how mechanical and electrical design is integrated, what structural works are required and so on.

Clients should investigate the CDM Co-ordinator's knowledge of the design process by asking them what experience they have had, asking for references from previous Clients and generally satisfying themselves that the person or organization they propose to appoint is conversant with the situations likely to be encountered.

CDM Co-ordinators must also have construction health and safety knowledge and will need to be able to demonstrate either general health and safety qualifications complied with construction site experience or specific construction health and safety qualifications.

12.6 Does the same level of competence have to be demonstrated for all projects?

No. HSE are at pains to point out that competency assessment programmes must be proportionate to the size and complexity of the project.

Knowledge needs to be relevant to the project size and future use, refurbishment and demolition of the structure.

The Client will be the best person to decide what level of competency they require for the CDM Co-ordinator and then, the CDM Co-ordinator can help the Client decide on competency levels for the design team, Principal Contractor and other specialist contractors.

12.7 Are there any specific guidelines issued by the HSE which would assist a Client in appointing a competent CDM Co-ordinator?

Yes. The CDM Regulations 2007 Approved Code of Practice contains a good competency framework for a CDM Co-ordinator, in Appendix 5.

The assessment of a CDM Co-ordinator should focus on the needs for the job and the CDM Co-ordinator may not need to demonstrate all of the competencies listed in the framework.

Generally, CDM Co-ordinators should be able to demonstrate the following.

12.7.1 Stage 1 of the Project

Task knowledge appropriate for the tasks to be undertaken.

Understanding of the design process.

Construction health and safety knowledge relevant to the tasks to be undertaken.

Check for membership of any professional bodies e.g. Chartered institution or professional institution.

Check for knowledge, experience and training in health and safety e.g. NEBOSH Construction Certificate, member of Institution of Occupational Safety and Health, member recognized health and safety register.

12.7.2 Stage 2 of the Project

Experience and ability sufficient to perform the task, to understand buildability, to recognize personal limitations, task related faults and errors and to be able to identify appropriate actions.

Experience relevant to the task.

Evidence of experience on similar projects, with comparable hazards, complexities and procurement route.

12.8 As a Client, we assessed the competency of the CDM Co-ordinator and decided they met our requirements and yet they failed to identify and take action on a significant hazard which resulted in an on-site accident. Will we be liable?

Every accident and incident will be investigated as appropriate by the HSE and they will be looking at the application of the CDM Regulations across the whole project, and in particular, in respect of the accident and the events leading up to it.

CDM Regulations 2007 do not expect every duty holder to have "perfect vision" and eliminate all risks from the construction project. Such as expectation would be unreasonable.

All duty holders are expected to be fully aware of their legal duties under CDM 2007 and to exercise them diligently. Everyone is expected to make judgements when assessing competence, determining hazard and risk, implementing the principles of prevention, etc.

Judgements are expected to be reasonable and to be taken based on sound interpretation of the facts presented, experience and understanding of the hazards and risks associated with the project, review of evidence, etc.

If a judgement can be proved to have been reasonable, based on the evidence presented, it is unlikely that action would be taken against the person making the judgement if the company or individual so appointed proved not to be competent to carry out the work.

12.9 The CDM Co-ordinator replaced the Planning Supervisor (as required under CDM 1994), are the competency requirements the same?

There will be many competencies which will be the same e.g. construction health and safety knowledge but the new CDM Co-ordinator role is seen to be more involved than the former Planning Supervisor role and there are many new duties which require the CDM Co-ordinator to actually do things as opposed to the previous "enabling role".

The CDM Co-ordinator must be in a position to advise the Client so he must have health and safety knowledge and experience as well as design process understanding.

Importantly, the new CDM Co-ordinator role requires evidence of good inter-personal skills whereas the Planning Supervisor role could previously have been carried out in significant isolation from the design team.

Someone carrying out the previous Planning Supervisor function may well have all of the competencies needed for the new CDM Co-ordinators role but it is not to be assumed. Clients will still need to assess the competency of their CDM Co-ordinators.

12.10 As a former Planning Supervisor, can I automatically become a CDM Co-ordinator?

Yes, for a while!

The transitory arrangements in the CDM Regulations 2007 enables any Planning Supervisor, appointed before April 2007, to automatically assume the role of CDM Co-ordinator until April 2008.

From April 2008, all duty holders will need to be able to demonstrate competence against the new competency framework as set out in the CDM ACOP.

Any former Planning Supervisor who fails to meet the competency requirements after April 2008 will not be able to take on the role of CDM Co-ordinator.

12.11 If a duty holder can only take on an appointment if they are competent, what happens if they are economical with the truth and take on something they should not?

Any duty holder who accepts an appointment for which he is not competent is liable to prosecution under Regulation 4 CDM 2007.

If their lack of competency has caused or contributed to a major accident or dangerous occurrence on the site then they may also be prosecuted under the Health and Safety at Work Etc Act 1974 and face a trial in the Crown Court where fines can be unlimited and custodial sentences awarded.

HSE Inspectors will look closely at the circumstances of the case and consider the management arrangements or the project, the role of the duty holders, etc.

Research has shown that a significant number of fatalities and major accidents in construction are caused by lack of competence, knowledge and training. Clients are appointing inexperienced professionals or contractors on economic grounds and such actions will not be tolerated.

12.12 How can the competency of organizations be assessed?

The competency of organizations should be carried out as a two stage process.

Stage 1   An assessment of the company's organization and arrangements for health and safety to determine whether they are sufficient to enable them to carry out the work safely and without risks to health.

Stage 2   An assessment of the company's experiences and track records to establish that it is capable of doing the work, understands the hazards and risks in so doing and has experience of the controls necessary to minimize risk.

There has been a lot of bureaucracy surrounding competency assessment schemes over the years and many of the processes adopted by organizations fail to address key health and safety issues and end up being a meaningless paper exercise.

Remember, the competency assessment scheme should be to provide the evidence needed that the person or organization to be appointed to the task is competent i.e. that they have knowledge and experience in carrying out the task.

Requiring unnecessary details and documents does not benefit anyone and certainly does not improve health and safety performance.

HSE have incorporated a set of "core competencies" into the CDM ACOP and it is hoped that some standardization of approach will reduce needless paperwork.

There are now a minimum of twelve core competencies which any organization can demonstrate – the detail needed to satisfy each competence will vary depending on the complexity of the project. Additional competencies will relate to CDM Co-ordinators, Principal Contractors and contractors and these will be determined by the Client e.g. evidence may be required of employer and public liability insurance.

No matter how much evidence is provided by the duty holder to demonstrate organization and arrangements for health and safety, the duty holder, especially the Client, should be asking for evidence, including references of previous, similar work experience.

12.13 Should Principal Contractors assess the competency of sub-contractors?

Yes. The duty to appoint competent persons is placed on every person and so the Principal Contractor will need to demonstrate that they have appointed competent and properly resourced contractors.

Many Principal Contractors already operate a scheme of approved contractor or nominated contractor. They will not need to do anything different except to evaluate their existing assessment scheme against the new "core criteria" to see of they are doing enough or whether they may need to amend their scheme.

A contractor may not need to be assessed for every project – those working in long term partnerships with main or Principal Contractors will not need to complete detailed questionnaires for each project – an annual assessment will suffice.

However, Principal Contractors should always ensure that their approved or nominated contractor has sufficient experience for the project in question and should ensure that any site-specific issues are addressed.

12.14 What are the requirements for assessing the competency of designers?

Designers will need to fulfil the same competency criteria as everyone else, namely:

- Capability to undertake the task
- Professional knowledge – demonstrated by membership of professional bodies
- Adequate resources

- Experience
- Knowledge of health and safety – or the access to competent advice.

Again, competency is relevant to the complexities of the project – a design company designing oil rigs for instance may need far greater knowledge of specialized health and safety subjects than a design company designing a small housing estate.

If a design company does not have, for instance, structural engineering knowledge, it will need to be able to demonstrate how it obtains that knowledge and how it ensures that the knowledge so obtained is competent i.e. the design company will have to show how it also assesses the competency of its specialist advisors and designers.

12.15 Is it necessary to assess the competency of individuals?

It is necessary to assess the competency of key individuals involved in a project as they will have an influential role to play in the success of the project.

A similar process of assessment to that of organizations can be undertaken i.e. stage 1 – individual qualifications and training and stage 2 – assessment of experience and track record.

Any person appointing another to a duty holders post must ensure that they are competent i.e. has the knowledge and experience to perform the task to the level expected.

All professionals will need to demonstrate that they have continuing professional development – often a recorded process by which attendance at seminars, conferences and training is recorded and externally verified by a professional body.

If continuing professional development is not relevant because the individual is not a member of one of the professions then they must demonstrate "life long learning" through attendance at seminars, reading journals, undertaking training, etc.

12.16 Are there any particular individual roles which should be assessed for competency?

Generally, because they have such an important part to play in ensuring good health and safety standards throughout the construction project, the following roles should all be assessed:

- Designer – everyone working in Senior roles
- CDM Co-ordinator – the specific person undertaking the duties
- Contracts Manager of the Principal Contractor
- Site Agent or Manager.

The competency of the Site Manager will be key in ensuring an overall culture of health and safety on the site – a Site Manager/Agent with a cavalier attitude to safety standards will convey the message that "anything will do" and operatives are likely to take short cuts and compromise safety whenever it seems beneficial.

12.17 A person we would like to appoint to the project team does not seem to have much experience, although they do have professional qualifications. What do we need to do to enable them to be appointed?

A person can be appointed to work on a construction project even though they may not meet all of the core competency criteria provided they work under the close supervision of someone who is.

We learn about our jobs from working with or watching more experienced colleagues and CDM 2007 recognizes this.

The critical aspect for ensuring competency and therefore safety, is to make sure that the person supervising the inexperienced operative is aware that they are responsible for ensuring that they work safely and for imparting their knowledge to them.

It would be sensible to introduce outcome based learning or formal reviews of the supervision the individual has had as this will create records to show that they have developed their knowledge and experience.

12.18 Are there any additional requirements regarding competency that the Principal Contractor should demonstrate?

The Principal Contractor should demonstrate competency relative to the construction project to be undertaken.

Any specialist aspects of the project should be clearly identified in the Pre-Construction Information Pack and for notifiable projects the CDM Co-ordinator should advise the Client of any specialist requirements relating to the Principal Contractor's appointment.

Not all Principal Contractors are competent in all areas of construction and a Client must not make such an assumption.

Additional requirements regarding competency for the following construction activities may be appropriate to the project:

• Demolition
• Working with hazardous substances

- Steel erection
- Temporary works.

Demolition contractors in particular will need to demonstrate competency, especially in relation to experience, as the demolition industry has the highest records of unsafe practices, fatalities and major injuries.

12.18.1 Demolition Accidents

- A 34-year-old demolition worker died when a 2 square metre section of wall fell on him during demolition works in a warehouse.
- The 4th and 5th floors of a five storey building undergoing renovation work collapsed trapping two workmen. The contractor was carrying out some internal demolition works as well as major refurbishment works. The incident could have had many casualties as debris from the collapsed floors dropped on to the busy street below.
- Four construction workers were killed when the entire structure they were demolishing collapsed unexpectedly trapping them all under tons of rubble. The contractor appointed normally carried out construction work and were not experienced in demolition works, but they thought that the demolition of the three storey house was straightforward and that "anyone could wield a sledgehammer". They took no advice from structural engineers, did not know what walls to prop up and did not plan the demolition. There were no risk assessments or method statements. The Client appointed them because they were the cheapest contractor for the new build part of the project and he assumed they would be able to demolish the building.

12.19 Does the Construction Skills Certificate Scheme Card demonstrate competency?

Generally, yes. The Construction Skills Certificate Scheme Card (CSCS) Scheme operates on two levels and the card is not awarded until an individual has demonstrated competency in both aspects.

Part 1 of the Scheme requires applicants to meet the trade or professional standards aspect and Part 2 is the health and safety competency aspect.

To demonstrate trade or professional competency an applicant has to produce evidence of qualifications and experience. Application forms have to be signed by employers or supervisors and must confirm that the individual has performed the trade/profession for x number of years, has qualifications and/ or membership of trade/professional bodies.

The health and safety competency aspect of the CSCS card is shown by completing a health and safety computer-based test. Various levels of competency are demonstrated through a range of card types e.g. visitor, operative, Manager, professional and the health and safety test is adapted to ensure a broad understanding of health and safety is demonstrated. Managers, Professionals,

Health and Safety Managers undertake the greatest number of multi-choice test questions and the pass rate is higher than the basic level card.

So anyone who has a CSCS card should be able to prove competency in their trade/profession and knowledge of health and safety.

Once an individual has a CSCS card they should need less comprehensive information, instruction and training on basic construction site safety. They will still need induction training on all new sites but it could be reasonably assumed that they have knowledge and understanding of the hazards and risks of say, working at heights.

13

The Health and Safety File

13.1 What is the Health and Safety File as required by the CDM Regulations 2007?

The Health and Safety File is required at the end of all notifiable projects.

The Health and Safety File provides information that will be needed by anyone who is preparing for construction work or cleaning work on an existing structure, including maintenance, repair, renovation, modification or demolition.

If prepared well, the Health and Safety File should be an invaluable document for all building owners and/or occupiers. It should contain information about the building which is relevant to health and safety.

There is no specific format to the Health and Safety File laid down by the Regulations, and it can therefore be in various forms e.g. paper file, drawings, on computer disk.

The Regulations do not specify the contents of the Health and Safety File except to require that it contains:

Information included with design by virtue of Regulation 18

and

Any other information relating to the project which it is reasonably foreseeable will be necessary to ensure the health and safety of any person at work who is carrying out or will carry out construction work or cleaning work in or on the structure or any person who may be affected by the work of such a person at work.

The Approved Code of Practice (L144) gives some guidance as to the content of the Health and Safety File as follows:

- Record or "as built" drawings and plans used and produced throughout the construction process along with design criteria
- General details of the construction methods and materials used
- Details of the structure's equipment and maintenance facilities
- Maintenance procedures and requirements for the structure
- Manuals produced by specialist Contractors and suppliers which outline operating and maintenance procedures and schedules for plant and equipment installed as part of the structure

- Details and location of utilities and services, including emergency and fire-fighting systems
- Residual hazards and risks within the structure e.g. location of hazardous substances, materials e.g. asbestos containing materials.

The information to be contained in the Health and Safety File should be agreed between the Client and the CDM Co-ordinator. The document, after all, should provide information for the building owner and they will have a valuable input into describing what information they believe will be relevant. Obviously, not all the Clients will know what to include in the Health and Safety File and in these instances the CDM Co-ordinator should give advice.

13.2 Who prepares the Health and Safety File and where does the information come from?

The CDM Co-ordinator has to prepare the Health and Safety File where one does not already exist, or must review and update any existing file.

This is a new specific duty for the CDM Co-ordinator under 2007 Regulations as previously, the CDM Co-ordinator only had to *ensure* that the File was prepared.

Information for the Health and Safety File comes from a variety of sources but key contributors are:

- Designers – the design risk assessments are important particularly where they indicate a "residual risk" associated with the design.
- CDM Co-ordinator – will have collected important information regarding previous uses of the site e.g. whether contaminated land is present, environmental hazards and so on.
- Principal Contractor – will have prepared construction sequences for the works, have details of materials and substances used.
- Structural Engineers – will have details of load bearing structures, imposed loadings on floors, guard rails, etc., details and location of foundations.
- Building Services Engineers – will have details of plant and equipment, operating and maintenance manuals, location of services, etc.
- Specialist Contractors e.g. Architectural Glaziers – will have details of types of glazing, fitting details, weight loadings, cleaning methods, etc.

It is the responsibility of the CDM Co-ordinator to co-ordinate the information to ensure that it is included in the Health and Safety File. They do not have to write the information although it would be sensible to compile an introduction and index to the information and detail of where such information is located if not all in one volume.

In many cases it can be extremely practical to have the Principal Contractor compile a draft Health and Safety File for the project as they see most of the information at first hand e.g. the amended drawings which show what was actually constructed and where, Operations and Maintenance Manuals from specialist Contractors, etc.

If the Principal Contractor compiles a draft Health and Safety File it must be handed over to the CDM Co-ordinator for checking, revisions and final drafting.

The agreement as to who is responsible for providing information for the Health and Safety File must be reached at the beginning of the project. Reference should be made in the Pre-Construction Information Pack as to who are the responsible persons, what the Health and Safety File is to contain and how many copies should be there and importantly, how soon after construction will it be compiled and completed.

13.3 What procedures should be followed to put the Health and Safety File together?

If put together properly the Health and Safety File should be one of the most beneficial requirements of the CDM Regulations.

How often have you taken ownership of a building (even a new house) and not had any idea where main services are located, whether certain materials have been used or how and when to maintain equipment and service plant? A good, detailed Health and Safety File should include all the information necessary to understand how, what, when, where, why a building is to be used and maintained safely.

A written procedure and checklist is a good starting point for compiling the Health and Safety File. The following steps could be considered:

- Define who will contribute information to the Health and Safety File.
- Agree with the Client what information they want in the File, how they want it compiled and how many copies are to be produced.
- Discuss with the Designers key information which will need to be included:
  - Design Risk Assessments
  - Details of residual risk
  - Specific construction methods
  - Structural details e.g. floor loadings.
- Agree the procedure for advising contractors to compile the File e.g. CDM Co-ordinator.
- Discuss with Building Services Consultants/Contractors key information in respect of services which will need to be provided e.g. Operations and Maintenance Manuals, how many and what format, drawings and so on.
- Advise the Designers and Principal Contractor of additional information which the Client has requested.
- Agree the procedure for site visits to obtain relevant information for the File during the construction phase.
- Obtain a list of all "Client Direct" appointments and establish what information they may have which will be relevant e.g. specialist installers.

- Write to all parties outlining the information you require them to provide and the timescales and deadlines you expect it in.
- Issue reminders through the Principal Contractor and Site Meeting minutes.
- Discuss with the Client formal procedures for withholding payment of accounts if information is not received.
- Visit site towards the end of the construction phase and conduct a Hazard and Risk Assessment of the premises to identify information which should be included in the File.
- Agree with the Client a handover procedure for the File, including a timescale as to when *realistically*, it will be available after the construction works are completed.

13.4 Who keeps the Health and Safety File?

The Client must retain a copy of the Health and Safety File. It must be delivered to him by the CDM Co-ordinator at the end of the construction works for each structure.

The Client has a duty under Regulation 10 to ensure that any information contained in the Health and Safety File is kept available for inspection by any person who may need information in the Health and Safety File for the purposes of complying with any statutory provisions.

The Client could delegate the responsibility for storing the Health and Safety File to another person e.g. the CDM Co-ordinator or Project Architect but the statutory duty to ensure that it is available will rest with the Client.

13.5 Where should the Health and Safety File be kept and how many copies of it should there be?

The Client has to retain a copy of the Health and Safety File in a format that can be easily used by other persons who may need the information.

The information in the Health and Safety File relates to health and safety issues and it is therefore essential that it is readily available, in both location and format, for people using, maintaining or cleaning the building or structure.

A copy of the Health and Safety File should be kept on the premises to which it relates so that it can be easily referenced by staff and maintenance personnel. However, site copies of documents have a tendency to be mislaid and if the Health and Safety File goes missing the Client will not be able to fulfil their duty under Regulation 10.

A practical solution is to prepare two Health and Safety Files – one a detailed Master Copy held at the Client's Head Office e.g. Property or Legal Department and a second summary version which contains essential health and safety information for day-to-day use and maintenance of the building within the premises itself.

The information in the Health and Safety File must be accessible to anyone who needs it. This will include Maintenance Contractors. They should be required to consult the Health and Safety File for methods of access to plant and equipment, maintenance procedures, potential hazards, etc. If a copy is not available for them on site and they undertake a task without being made aware of any residual hazards or procedures to be taken and subsequently have an accident, the resulting accident investigation could conclude that had they had access to the information the accident would not have occurred, and the Client could be charged with contravening the Regulations. The maintenance company could then sue for negligence in the Civil Courts.

The location, format and numbers of the Health and Safety File should be discussed and agreed between the CDM Co-ordinator and the Client at the outset of the project. The resulting decisions should be recorded in the Pre-Construction Information Pack.

13.6 With whom does liability for the Health and Safety File rest?

The Client is legally liable for ensuring that the Health and Safety File is available to any person who may need the information contained within it in order to comply with relevant statutory provisions.

The Client is legally liable for ensuring that the File is transferred to the structure's new owners on disposal of his interests in the "property of the structure".

The Client is legally liable for ensuring that leaseholders of any parts of the building have access to the information contained within the Health and Safety File. This may include issuing all leaseholders with a copy but this is not mandatory.

The CDM Co-ordinator is legally liable for preparing the Health and Safety File.

The CDM Co-ordinator is also legally liable for ensuring that the Health and Safety File is "reviewed, amended or added to" in order to ensure that the information referred to in the Regulations is contained within the File.

The CDM Co-ordinator is legally liable for ensuring that, on completion of the construction work on each structure comprised in the project, the Health and Safety File is delivered to the Client.

The duty placed on the CDM Co-ordinator will be "as far as is reasonably practicable". It would be unreasonable and impractical to expect a CDM Co-ordinator to be fully conversant with the detailed knowledge necessary to approve specialist Contractor information. What would be reasonable is to expect the CDM Co-ordinator to ask the specialist critical questions in respect of health and safety e.g.

- Has safe access been designed
- How is access to be gained

- Have any fragile materials been used and if so, where
- What hazardous materials/substances have been used, and where
- What residual risks remain
- Have safety notices been displayed
- Have any hazardous areas been created e.g. confined spaces
- What health and safety management systems are recommended to mitigate residual risk
- What personal protective equipment is expected to be used
- How and when is maintenance to be carried out
- How and when is cleaning to be carried out.

The CDM Co-ordinator should be skilled at picking out unusual and key site-specific health and safety issues and *must* ensure that relevant information is included in the File. A competent health and safety professional does not have to have detailed knowledge of every job task to be able to identify common hazards and risks and to apply the principles of the hierarchy of risk control to the process.

13.7 Once the Health and Safety File has been completed and handed over to the Client, who is responsible for keeping it up-to-date?

The Client retains responsibility for the Health and Safety File once it is handed over to him.

The File will need to be handed over to future CDM Co-ordinators and any other contractors whenever the Client commissions works which fall within the jurisdiction of the CDM Regulations.

It will be the duty of the CDM Co-ordinator to keep the File amended, reviewed and added to in respect of all new works. When these works are complete the File will be handed back to the Client in its amended form.

If subsequent construction works are not notifiable to the HSE, there will be no duty on the Client to update and amend any existing Health and Safety File, although, of course, it would be good practice to do so.

Clients need to be mindful of the fact that Health and Safety Files may not be updated when minor works are carried out and that they could feasibly, give contractors and designers inadequate information e.g. an original Health and Safety File included references to encapsulated asbestos material yet subsequent works may have had the asbestos removed. If the contractor had the old Health and Safety File he may make unnecessary plans to deal with and manage the risk of asbestos on site.

Whenever a Health and Safety File is provided for a building the Client or subsequent building owner/occupier should develop a robust procedure for ensuring it is kept up-to-date. This will help reduce site-safety hazards caused by lack of knowledge, etc.

There is a growing tendency as buildings change hands for the legal profession to require existing owners to confirm in writing that the Health and Safety File is available and up-to-date. If a subsequent owner found it to be deficient they could instigate civil proceedings. Equally, both freeholders and mortgage lenders e.g. commercial banks are making the Health and Safety File a legal item in conveyancing and lease-drafting work.

However, having gone to the not inconsiderable expense of having a Health and Safety File prepared, it seems foolish not to ensure that it is kept up-to-date as improvement and repair works are undertaken within the building.

13.8 What happens if certain information which should be included in the Health and Safety File is not available?

The information required for the Health and Safety File should be made clear at the beginning of the project, including who is responsible for providing what.

If all the information necessary to complete the File is not available at the end of the construction works the CDM Co-ordinator may have to provide an incomplete Health and Safety File to the Client, with details of the outstanding information.

The information which is required for the File is that which is "reasonably foreseeable will be necessary to ensure the health and safety of any person at work . . .". Certain information may *not* have been reasonably foreseeable during the project and may only be collected retrospectively. This information could reasonably be expected to be excluded from the Health and Safety File when first compiled but the CDM Co-ordinator would be expected to "add to" the File by providing the information as soon as it is available.

Contractors and Client Direct appointments who fail to provide relevant information as requested should be held in breach of contract and financial penalties should be imposed e.g. retention fees increased to 20% of contract value for instance.

The CDM Co-ordinator should do all that is reasonably practicable to obtain the information e.g. contacting Contractors directly and keep records to that effect.

Those which fail to comply with the requirements to provide information should be judged as lacking either competency or resources to fulfil their obligations under CDM. The Client should be informed of such instances and advice given to "de-list" the Contractor, etc. The Client could write to the Contractor/Company advising them that they will be de-listed.

Contractors will be in breach of Regulation 19 if they fail to provide information promptly to the Principal Contractor. The Principal Contractor will be in breach of Regulation 22 if they fail to provide information promptly to the CDM Co-ordinator.

Prima facie evidence of failing to comply knowingly with a legal duty should imply lack of competency and should the Client continue to appoint a Contractor or Designer in those circumstances, the Client could be held to be in breach of Regulation 14.

Case Study

The Electrical Contractor on a project went into liquidation shortly after the completion of the construction works on a new restaurant project, and before the Electrical Operations and Maintenance Manual had been completed and handed over to the CDM Co-ordinator.

A major, though not fatal, electrical accident occurred to a staff member and following the reporting of the accident, the local Environmental Health Officer conducted an Investigation. He asked for Electrical Test Certificates to verify correct installation of the works under the Electricity at Work Regulations 1989. Those Certificates could not be produced. The EHO shut down the entire kitchen operation until such time as he could be satisfied that the electrics were safe. The Client commissioned another electrical Contractor to undertake a full test and survey of all of the electrics and to produce the appropriate Test Certificates, together with a suitable Operations and Maintenance Manual. When the subsequent test showed all the electrical wiring installation to be satisfactory the Environmental Health Officer lifted the Prohibition Notice and the kitchen re-opened. The suspect piece of kitchen equipment was seized for examination.

Had the original Test Certificates been readily available, the Client would have been able to demonstrate immediately that the electrical installation met all the safety criteria. As the documents were not available valuable trade was lost for 3 days – the time taken to organize re-testing and commissioning of all electrical equipment and installations within the premises.

The Client and the CDM Co-ordinator reviewed procedures for the handing over of key information at the completion of works and agreed that on future projects, unless a copy of the Electrical Installation Test Certificate was provided at the handover meeting or was faxed to the CDM Co-ordinator prior to the handover meeting, project completion would not be achieved and liquidated damages and other contractual financial penalties would be imposed. The Principal Contractor was held responsible for co-ordinating the information from the Electrical Contractor.

Designers will not be able to design and specify any aspect of the building which might contravene the Regulations.

It could be assumed that the Designers always designed buildings in accordance with the Regulations but unfortunately, there are many accidents to building users, or maintenance contractors, which are caused by poor design or poor use of materials.

Examples could be:

- Unnecessary changes in floor level which create steps which are so shallow that they are not easily noticeable, causing trip hazards.
- Upper floor balconies with no upstand, enabling debris to be kicked under the bottom rail, falling on to people below.
- Poor specification of floor coverings which do not have the correct "slip co-efficient" causing people to slip frequently.
- Installation of large expanses of translucent glazing causing people to walk in to it because they do not notice that it is there.
- Poor access to plant and equipment.
- Inadequate ventilation to work rooms.
- Pedestrian and vehicle routes without adequate separation.

14.3 Do Clients have any responsibility to ensure that the Designers design a workplace in accordance with the Workplace (Health, Safety and Welfare) Regulations 1992?

Yes.

Clients have to ensure that any person with duties under CDM 2007 takes all such reasonable steps regarding the management of health and safety to ensure compliance with the Workplace (Health, Safety and Welfare) Regulations 1992.

As Designers have duties under CDM 2007, the Client must make sure that there are procedures in place to consider the Regulations and that the Designers comply with the legal requirements.

Clients do not have to check the Designers designs and specification, although they can ask their CDM Co-ordinator (on notifiable projects) to check these for them.

Client should, however, require the Designers to sign a declaration that they have met the requirements of the Workplace Regulations.

Clients may want to set out specific performance criteria for the building or materials e.g. the type of floor covering to be used in certain areas so that the Designer is provided with appropriate use of the building and can adapt their designs accordingly.

14.4 Are Principal Contractors covered by the same duty to ensure that structures comply with the Workplace (Health, Safety and Welfare) Regulations 1992?

Yes.

If a Principal Contractor amends a design or specifies a material for a structure to be used as a workplace then the Client must ensure that the Principal Contractor complies with the requirements of the Workplace (Health, Safety and Welfare) Regulations 1992.

Principal Contractors, and even sub-contractors or specialist contractors, will often make last-minute changes to designs so as to enable the structure or feature to be constructed. Whilst the design change may make construction easier, it might make future maintenance of the building or its future use more difficult and cause a contravention of a specific regulation.

All parties to the design and construction process must be familiar with the Workplace (Health, Safety and Welfare) Regulations 1992.

14.5 What are some of the specific requirements of the Regulations?

14.5.1 Health issues

Ventilation

Workplaces need to be adequately ventilated. Fresh, clean air should be drawn from a source outside the workplace, uncontaminated by discharges from flues, chimneys or other process outlets, and be circulated through the workrooms.

Ventilation should also remove and dilute warm, humid air and provide air movement which gives a sense of freshness without causing a draught. If the workplace contains process or heating equipment or other sources of dust, fumes or vapours, more fresh air will be necessary to provide adequate ventilation.

Windows or other openings may provide sufficient ventilation but, where necessary, mechanical ventilation systems should be provided and regularly maintained.

These Regulations do not prevent the use of unflued heating systems designed and installed to be used without a conventional flue.

Any mechanical ventilation, including air conditioning systems, should be regularly and properly cleaned, tested and maintained to ensure that they are kept clean and free from anything which may contaminate the air.

If ventilations systems are required to provide dilution for obnoxious fumes, etc. they may need a breakdown warming device fitted.

14.5.2 Safety issues

Maintenance

The workplace and certain equipment, devices and systems should be maintained in efficient working order (efficient for health, safety and welfare). Such maintenance is required for mechanical ventilation systems, equipment and devices which would cause a risk to health and safety or welfare should a fault occur.

Floors and traffic routes

"Traffic route" means a route for pedestrian traffic, vehicles, or both, and includes any stairs, fixed ladder, doorway, gateway, loading bay or ramp. There should be sufficient traffic routes, of sufficient width and headroom, to allow people and vehicles to circulate safely with ease.

Floors and traffic routes should be sound and strong enough for the loads placed on them and the traffic expected to use them. The surfaces should not have holes, be uneven or slippery and should be kept free of obstructions.

Restrictions should be clearly indicated. Where sharp or blind bends are unavoidable or vehicles need to reverse, measures such as one-way systems and visibility mirrors should be considered. Speed limits should be set. Screen should be provided to protect people who work where they would be at risk from exhaust fumes, or to protect people from materials likely to fall from vehicles.

Additional measures need to be taken where pedestrians have to cross or share vehicle routes. These may include marking of routes, provision of crossing points, bridges, subways and barriers.

Open sides of staircases should be fenced with an upper rail at 900 millimetre or higher and a lower rail. A handrail should be provided on at least one side of every staircase and on both sides if there is a particular risk. Additional handrails may be required down the centre of wide staircases. Access between floors should not be by ladders or steep stairs.

Falls into dangerous substances

Falling into dangerous substances can have devastating consequences and often be a cause of deaths in the workplace.

High standards of protection are therefore required to tanks, pits or other structures.

Any tank, pit or other structure where there is a risk of people falling must be covered or securely fenced.

Traffic routes passing any structures or areas in which people could fall into dangerous substances must be fenced.

Transparent and translucent doors, gates or walls and windows

Windows, transparent or translucent surfaces in walls, partitions, door and gates should, where necessary for reasons of health and safety, be made of safety material or be protected against breakage. If there is a danger of people coming into contact with it, it should be marked or incorporate features to make this apparent.

Employers will need to consider whether there is a foreseeable risk of people coming into contact with glazing and being hurt. If this is the case, the glazing will need to meet the requirements of the Regulations.

Openable windows and the ability to clean them safely

Openable windows, skylights and ventilators should be capable of being opened, closed or adjusted safely and when open, should not be dangerous.

Windows and skylights should be designed so that they may be cleaned safely. When considering if they can be cleaned safely, account may be taken of equipment used in conjunction with the window or skylight or of devices fitted to the building.

Doors and gates

Doors and gates should be suitably constructed and fitted with safety devices (e.g. panic bolts (a door bolt that can be operated from the inside in an emergency)), if necessary.

Doors and gates which swing both ways and conventionally hinged doors on main traffic routes should have a transparent viewing panel.

Power-operated doors and gates should have safety features to prevent people being struck or trapped and, where necessary, should have a readily identifiable and accessible control switch or device so that they can be stopped quickly in an emergency.

Upward-opening doors or gates need to be fitted with an effective device to prevent them falling back. Provided that they are properly maintained, counterbalance springs and similar counterbalance or ratchet devices to hold them in the open position are acceptable.

Escalators and moving walkways

Escalators and moving walkways should function safely, be equipped with any necessary safety devices, and be fitted with one or more emergency stop controls which are easily identifiable and readily accessible.

14.5.3 Welfare issues

Sanitary conveniences and washing facilities

Suitable and sufficient sanitary conveniences and washing facilities should be provided at readily accessible places. They and the rooms containing them should be kept clean and be adequately ventilated and lit. Washing facilities should have running hot and cold or warm water, soap and clean towels or other means of cleaning and drying. If required by the type of work, showers should also be available. Men and women should have separate facilities unless each facility is in a separate room with a lockable door and is for use by one person at a time.

Drinking water

An adequate supply of wholesome drinking water, with an upward jet or suitable cups should be provided. Water should only be provided in refillable enclosed containers where it cannot be obtained directly from a mains supply. The containers should be refilled at least daily (unless the are chilled water dispensers where the containers are returned to the supplier for refilling). Bottled water/water dispensing systems may still be provided as a secondary source of drinking water.

Accommodation for clothing and facilities for changing

Adequate, suitable and secure space should be provided to store worker's own clothing and special clothing. As far as is reasonable practicable, the facilities should allow for drying clothes. Changing facilities should also be provided for workers who change into special work clothing. The facilities should be readily accessible from workrooms and washing and eating facilities and should ensure the privacy of the user.

Facilities for rest and to eat meals

Suitable, sufficient and readily accessible rest facilities should be provided. Rest areas or rooms should be large enough and have sufficient seats with backrests and tables for the number of workers likely to use them at any time. They should include suitable facilities to eat meals where meals are regularly eaten in the workplace and the food would otherwise be likely to be contaminated.

Seats should be provided for employees to use during their breaks. These should be in a place where personal protective equipment need not be worn. Work areas can be counted as rest areas and as eating facilities providing they are adequately clean and there is a suitable surface on which to place food. Where provided eating facilities should include a facility for preparing or obtaining a hot drink. Where hot food cannot be obtained in or reasonable near to the workplace, employees may need to be provided with a means for heating their own food.

Canteens or restaurants may be used as rest facilities provided there is no obligation to purchase food.

Suitable rest facilities should be provided for pregnant women and nursing mothers. They should be near to sanitary facilities and, where necessary, include the facility to lie down.

Rest areas should be away from the workstation and should include suitable arrangements to protect non-smokers from discomfort caused by tobacco smoke.

14.6 Does an employer have to legally provide employees with showers?

No, unless the nature of the work requires them, or they are needed for health reasons.

Employees who are exposed to dusty atmospheres may require showers to ensure that they decontaminate themselves before leaving the workplace.

Asbestos workers are required by law to be given showering facilities (under the Control of Asbestos Regulations 2006).

Catering staff or kitchen workers may need to have showers provided because of the hot, humid atmosphere of the kitchen or because showering helps to reduce surface body bacteria which in turn will reduce the risk of food contamination.

Remember that the law is, however, a minimum set of standards, and additional welfare and washing facilities may prove a good investment in respect of employee morale, etc.

14.7 What are the recommended numbers of water closets, urinals and wash hand basins which need to be provided for employees?

The "Workplace" Regulations only stipulate that "suitable and sufficient" facilities should be provided.

The guidance in the Approved Code of Practice gives and indication of suitable and sufficient facilities as:

No. of people at work (Male and Female)	No. of water closets	No. of wash stations
1–5	1	1
6–25	2	2
26–50	3	3
51–75	4	4
76–100	5	5

If accommodation is to be provided for men only, then water closets and urinals can be provided as follows:

No. of men at work	No. of water closets	No. of urinals
1–15	1	1
16–30	2	1
31–45	2	2
46–60	3	2
61–75	3	3
76–90	4	3
91–100	4	4

Wash hand basins should be provided for every water closet plus reasonable additions for the urinals.

Checklist

The Workplace (Health, Safety and Welfare) Regulations 1992

All places of work where employees work must comply with applicable requirements e.g.:

- Workplaces to be kept clean, well maintained, in good order and repair. Equipment or premises which could fail must be subject to adequate maintenance procedures.
- Ventilation must be suitable and sufficient, incorporating fresh or purified air. Any mechanical ventilation systems must be adequately maintained.
- Reasonable temperatures must be maintained during working hours inside all buildings. No injurious or offensive fumes may enter the workplace. Suitable thermometers to be provided. Recommended temperature for sedentary work is 16°C *but* it is only advisory.
- Suitable and sufficient lighting is required and where artificial light is provided, if it should fail, emergency lighting is required.
- Walls, floors and ceiling surfaces must be capable of being kept clean. Furniture and fixtures, etc. must be kept clean.
- Persons at work to have adequate working space and must not be overcrowded.
- Workstations must be suitably positioned, and if outside, suitably protected from the weather.
- Arrangements must be in place for leaving the workstation in an emergency.
- Floors and traffic routes must be constructed suitable for the purpose i.e. having no holes or slopes.
- Windows, doors and any panels, etc. glazed wholly or partially with translucent glass must be of safe material and adequately marked. Risk assessment on glazing is required.
- Windows and skylights should be designed so that when opened, closed or adjusted they cause no danger to anyone.
- Safe circulation of pedestrians and vehicles must be organized in the workplace. Traffic routes are required.
- Doors and gates must be fitted with any necessary safety devices e.g. to sliding doors.
- Escalators must function safely.
- Sanitary accommodation must be suitable and sufficient and readily accessible to all persons at work.

- Washing facilities must be provided together with adequate soap, towels, hot water, etc.
- Wholesome drinking water must be provided.
- Suitable accommodation for outdoor clothing is required, and suitable changing facilities must be provided.
- Rest facilities are to be provided, including facilities to eat meals and take breaks.
- Suitable facilities must be provided for pregnant women or nursing mothers.

15

Management of Health and Safety on Construction Projects

15.1 What aspects of the Management of Health and Safety at Work Regulations 1999 apply to construction sites?

The Management of Health and Safety at Work Regulations 1999 (MHSWR) set out general duties of employers and employees in all non-domestic work activities and aim to improve health and safety management by developing the general principles set out in the Health and Safety at Work Etc Act 1974.

MHSWR duties overlap with duties contained in several other pieces of health and safety management by developing the general principles set out in the Health and Safety at Work Etc Act 1974.

MHSWR duties overlap with duties contained in several other pieces of health and safety, including the CDM Regulations. Compliance with other legislation normally implies compliance with MHSWR, but sometimes the duties in MHSWR go beyond those of other regulations. In these instances, the duties imposed by MHSWR take precedence over others.

MHSWR places duties on employers (and the self-employed) including Clients, designers, CDM Co-ordinators, principal contractors and other contractors.

Under MHSWR employers must:

- Assess the risks to the health and safety of their employees and others who may be affected by the work activity (Regulation 3).
- Identify what actions are necessary to eliminate or reduce the risks to health and safety of their employees and others.
- Apply the principles of prevention and protection.
- Carry out and record in writing, if they have five employees or more, a risk assessment.
- Make appropriate arrangements for managing health and safety, including planning, organization, control, monitoring and review of preventative and protective measures. Arrangements must be recorded if five or more employees.
- Provide appropriate health surveillance for employees whenever the risk assessment shows that it is necessary e.g. to check for skin dermatitis.

- Appoint competent persons to assist with the measures needed to comply with health and safety laws. Competent persons should ideally be from within the employer's own organization. Where more than one competent person is appointed the employer must ensure that adequate co-operation exists between them.
- Set up procedures to deal with emergencies and liaise, if necessary, with medical and rescue services.
- Provide employees with relevant information on health and safety in an understandable form.
- Co-operate with other employees sharing a common workplace and co-ordinate preventative and protective measures for the benefit of all employees and others.
- Make sure that employees are not given tasks beyond their capabilities and competence.
- Ensure that employees are given suitable training.
- Ensure that any temporary workers are provided with relevant health and safety information in order to carry out their work safely.

Employers have duties under MHSWR to:

- Use equipment in accordance with training and instruction.
- Report dangerous situations.
- Report any shortcomings in health and safety arrangements.
- Take reasonable care of their own and other's health and safety.

The Principal Contractor will carry the bulk of the responsibility for MHSWR on a construction site and as the site will be "multi-occupied", the Principal Contractor must ensure co-operation and co-ordination between employers. This will be laid out in the Construction Phase Health and Safety Plan. Contractors must carry out their own risk assessments but the Principal Contractor must complete these where hazards and risks affect the whole workforce e.g. site access routes, communal lifting operations. As a multi-occupied site, the Principal Contractor will assume overall responsibility for the management of health and safety and will co-ordinate and arrange emergency procedures, etc. Information on such procedures must be given to all persons using the site by the Principal Contractor. Information must be comprehensible and understandable so may need to be in picture form, carbons, posters and foreign language, etc.

15.2 What are the five steps to successful health and safety management?

The five steps to successful health and safety management are exactly the same steps required for the successful management of any project, namely:

(1) Policy
(2) Organize

(3)   Planning and implementing
(4)   Measuring performance
(5)   Reviewing performance.

15.2.1 Step 1: set your own policy

All employers who have employees must have a Health and Safety policy and those who have five or more employees must have it written down and available for employees to consult.

All main contractors will have a Health and Safety Policy and this sets the framework for the management of health and safety on any site on which their employees are to work.

The Site Agent, Manager or Contracts Manager should review the Company Policy and see whether amendments or additions are necessary for the actual site in question. Does the main Policy reflect how you will manage health and safety on site? Have you identified your own safety objectives?

15.2.2 Step 2: organize

Having identified the overall Health and Safety Policy and objectives for the site, the site and its operatives need to be organized to delivery the Policy and objectives. You will need to create a positive health and safety culture for the site – setting standards, enforcing standards, taking a strong lead, etc.

There are four components to a health and safety culture:

(1)   Competence – recruitment, training and advisory support
(2)   Control – allocating responsibilities, securing commitment, instruction and supervision
(3)   Co-operation – between individuals, other contractors and employers
(4)   Communication – spoken, written, visible.

In organizing the site Health and Safety Policy and procedures have you:

- Allocated responsibility for health and safety to specific people? Are they clear on what they have to do and are they held accountable?
- Consulted and involved all operatives, contractors, sub-contractors, the self-employed and other employers, trade union representatives, etc.?
- Given everyone sufficient information on the health and safety standards, objectives, hazards and risks of the site?
- The right level of expertise on site to manage all tasks safely and effectively?
- A properly trained workforce and fully inducted operatives?
- Specialist advice available to assist you in managing health and safety?

All of the above should be considered for each construction site before the work actually commences.

15.2.3 Step 3: Plan and set standards

If the CDM Regulations 2007 Part 3 apply to the construction project there must be a Construction Phase Health and Safety Plan before works start on site. Even if the project is not notifiable under CDM 2007 it is still a requirement to formulate a Construction Phase Health and Safety Plan which is proportionate to the risks associated with the project.

What objectives do you want for the project?

Has the Client set these? Under CDM 2007 the Client has a greater role in setting health and safety standards for projects. Some such standards could be:

- No major injury accidents
- Lost working days to be no more than 1% of construction days
- Risk Assessments to be submitted 24 hours before works commence
- All operatives to be inducted to the Site-Safety Standard within the first hour of commencement on day one.
- No visitors to be let into the construction area unaccompanied.
- All persons to wear as a minimum:
  - Safety helmet
  - Safety shoes
  - Hi-vi vest.

Set some achievable objectives. Decide how you are going to monitor that you have achieved them. For example how will you require all accidents to be recorded – or will you only require reportable accidents to be notified to you.

What mechanism will you introduce so that health and safety can be considered before any major work activity or site alteration.

What plans have you put in place for emergency procedures?

15.2.4 Step 4: Measure Performance

Consider how you will monitor the performance of health and safety controls and standards.

Who will carry out safety checks – how often?

Will there be a sub-contractors safety meeting?

Who will review contractors' Risk Assessments and Method Statements?

How will accident statistics be reviewed?

Who will worry about near misses?

Will the number of HSE inspections be recorded? Is there a system?

What is the lost time rate caused by accidents and ill-health?

Can any costs be quantified for lost time due to work stopping to investigate an accident?

What are the standards required by the Client and how well do you meet them e.g. Permit to Work procedures?

15.2.5 Step 5: Audit and Review

How well does the Health and Safety Plan work? Where can it be improved?
How do you learn from mistakes and successes?
If on-site-safety checks are carried out, who implements action plans?
Can you review performance on this project against other projects?
Can you benchmark your health and safety compliance with competitors.

Health and safety management is a journey of continuous improvement, never be satisfied with "making do" – the better the health and safety management, the safer the site and the more efficient and profitable the job. And the greater the contractor's reputation.

15.3 Does the Principal Contractor have to carry out site-safety audits?

The Principal Contractor has a duty to plan, manage and monitor the construction phase in a way which ensures that, so far as is reasonably practicable, it is carried out without risk to the health and safety of persons working on the site or who may be affected by the works.

Also, the Principal Contractor has a duty to ensure that the principles of prevention are followed.

In order to ensure that the construction phase is monitored, the Principal Contractor would really have to carry out some kind of site-safety audit or check, or would need to ensure that others complete the checks.

Many Principal Contractors will monitor standards themselves by carrying out site-safety checks as part of their own management standards, or they will employ third party safety inspectors/auditors to visit site and carry out independent checks.

There is no specific legal requirement to carry out checks, but a Principal Contractor may have difficulty demonstrating "due diligence" if things go wrong on site i.e. he will have no records to prove that he has a mechanism in place for monitoring site safety.

An alternative way for the Principal Contractor to monitor site-safety standards is to require all sub-contractors to carry out their own site-safety audits and forward checklists to him for review, action and filing.

Site-safety audit results could be discussed at the site meetings held with all contractors on site and could be part of the communication process and worker involvement in health and safety.

15.4 Does the Client have to carry out site-safety audits?

No, not specifically.

The Client has to ensure that management arrangements are in place (for health and safety) during the course of the project and must ensure that there is a process for reviewing those arrangements during the project.

Some Clients will require the CDM Co-ordinator, on notifiable projects, to monitor and review site-safety conditions and issue a report to all relevant parties. In this way the Client will be re-assured that the Principal Contractor is doing what they should do.

Any procedure or record-keeping system which shows that a Client was aware of his legal responsibilities and had a system in place to ensure that they were fulfilled would be invaluable should any situation lead to a court appearance or involve enforcement activity.

If the Client didn't want to carry out site-safety audits himself or commission others to do so on his behalf, he could ensure that a health and safety item was included on the site minutes and that a regular update of key issues relating to health and safety was issued.

15.5 What are the requirements to manage health and safety under the CDM Regulations 2007?

Every Client shall take reasonable steps to ensure that the arrangements made for managing the project, including the allocation of sufficient time and resources, by persons with a duty under the Regulations are suitable to ensure that:

- Construction work can be carried out so far as is reasonable practicable without risks to the health and safety of any person.
- The requirements regarding site-welfare facilities, as listed in Schedule 2 of the Regulations, are complied with in respect of any person carrying out construction work.
- Any place designed for use as a workplace has been designed taking into account the Workplace, (Health, Safety and Welfare) Regulations 1992 – as far as those Regulations relate to the design of, and materials used in, the structure.

The Client must also take reasonable steps to ensure that the arrangements are maintained and reviewed throughout the project.

The Client must be satisfied that any contractor appointed to a project has made arrangements for managing the project.

The Management of Health and Safety at Work Regulations 1999 would provide a good guide to Clients – or anyone else – on the steps which need to be taken to manage health and safety.

The Client does not need to be expert in health and safety nor construction processes but know enough to know that a considered approach has been taken.

Minimum management standards would be:

- Site Agent appointed
- Construction Phase Health and Safety Plan
- Programme for appointing competent contractors

- Site specific hazard identification
- Risk Assessments and Method Statements
- PPE policy
- Provision of welfare facilities
- Procedures for communication across all those involved in the project e.g. regular site progress meetings
- Accident management procedures
- First Aid provision.

Reasonable steps have to be taken to check that management arrangements are in place, maintained and reviewed.

This does not mean that the Client has to go to site every day or week but it could be that the Client instructs his Project Manager or Designer to note health and safety management arrangements at the site meetings.

On notifiable projects, the CDM Co-ordinator could visit site and report on facilities and procedures – such a visit could be combined with obtaining information for the Health and Safety file.

The CDM Regulations 2007 do not define "reasonable steps", nor does the Approved Code of Practice. General interpretation of "reasonable" is that as long as the arrangements put in place focus on the needs of the particular job and are proportionate to the risks arising from the work, then the arrangements will be reasonable.

15.6 What are the general management duties which apply to construction projects?

The Management of Health and Safety at Work Regualtions (MHSWR) set out general duties of employers and employees in all non-domestic work activities and aim to improve health and safety management by developing the general principles set out in the Health and Safety at Work Etc Act 1974.

MHSWR duties overlap with duties contained in several other pieces of health and safety legislation, including the CDM Regulations. Compliance with other legislation normally implies compliance with MHSWR, but sometimes the duties in MHSWR go beyond those of other regulations. In these instances, the duties imposed by MHSWR take precedence over others.

MHSWR places duties on employers (and the self-employed) including clients, designers, CDM Co-ordinators, principal contractors and other contractors.

Under MHSWR employers must:

- Assess the risks to the health and safety of their employees and others who may be affected by the work activity (Regulation 3).
- Identify what actions are necessary to eliminate or reduce the risks to health and safety of their employees and others.
- Apply the principles of prevention and protection.
- Carry out and record in writing, if they have five employees or more, a risk assessment.

- Make appropriate arrangements for managing health and safety, including planning, organization, control, monitoring and review of preventative and protective measures. Arrangements must be recorded if five or more employees.
- Provide appropriate health surveillance for employees whenever the risk assessment shows it necessary e.g. to check for skin dermatitis.
- Appoint competent persons to assist with the measures needed to comply with health and safety laws. Competent persons should ideally be from within the employer's own organization. Where more than one competent person is appointed the employer must ensure that adequate co-operation exists between them.
- Set up procedures to deal with emergencies and liaise, if necessary, with medical and rescue services.
- Provide employees with relevant information on health and safety in an understandable form.
- Co-operate with other employees sharing a common workplace and co-ordinate preventative and protective measures for the benefit of all employees and others.
- Make sure that employees are not given tasks beyond their capabilities and competence.
- Ensure that employees are given suitable training.
- Ensure that any temporary workers are provided with relevant health and safety information in order to carry out their work safely.

Employees have duties under MHSWR to:

- Use equipment in accordance with training and instruction
- Report dangerous situations
- Report any shortcomings in health and safety arrangements
- Take reasonable care of their own and other's health and safety.

The Principal Contractor will carry the bulk of the responsibility for MHSWR on a construction site and as the site will be "multi-occupied", the Principal Contractor must ensure co-operation and co-ordination between employers. This will be laid out in the Construction Phase Health and Safety Plan. Contractors must carry out their own risk assessments but the Principal Contractor must complete these where hazards and risks affect the whole workforce e.g. site access routes, communal lifting operations. As a multi-occupied site, the Principal Contractor will assume overall responsibility for the management of health and safety and will co-ordinate and arrange emergency procedures, etc. Information on such procedures must be given to all persons using the site by the Principal Contractor. Information must be comprehensible and understandable so may need to be in picture form, carbons, posters and foreign language, etc.

The CDM Regulations 2007 strengthen the statutory duties on all persons regarding co-operation, co-ordination and communication. These are the three key principles underpinning good health and safety management.

16

Accident and Incident Management

16.1 What are the requirements of the Report of Injuries Diseases and Dangerous Occurrence Regulations 1995?

Where any person dies or suffers any of the injuries or conditions specified in Appendix 1 of the Regulations, or where there is a "dangerous occurrence" as specified in Appendix 2 of the Regulations, as a result of work activities, the "responsible person" must **notify the relevant enforcing authority.**

Notification must be by telephone or fax and confirmed in writing within 10 days.

Where any person suffers an injury not specified in the Appendix but which results in an absence from work of more than 3 calendar days the "responsible person" must notify the enforcing authority in writing.

The "responsible person" may be the employer, the self-employed, someone in control of the premises where work is carried out or someone who provided training for employment.

Where death of any person results within 1 year of any notifiable work accident the employer must inform the relevant enforcing authority.

When reporting injuries, diseases (e.g. industrial diseases (contracted as a result of work undertaken e.g. Weil's Disease, Miner's Lung)) or dangerous occurrences the approved forms must be used, either F2508 or F2508a.

Records of all injuries, diseases and occurrences which require reporting must be kept for at least 3 years from the date they were made.

Accidents to members of the public which result in them being taken to hospital as a result of the work activity must be reported.

Incidents of violence to employees which result in injury or absence from work must be reported.

16.2 Why must these type of accidents be reported?

National accident statistics are collated by the Health and Safety Executive (HSE) in order to indicate the general state of health and safety across the country. Fatalities, major injuries and "over three day" injuries are all recorded and allocated to industry specific sectors so that the state of legal compliance, accident trends, etc. can be judged.

However, the most important reasons for notifying accidents are:

- Its a legal requirement
- The enforcing authorities can investigate to establish whether the employer has contravened the law
- Serious incidents can be prevented from happening again.

Accident statistics for 2000/2001 are:

- 249 fatalities
- 27,477 major injuries
- 127,084 "over three day" injuries
- 384 fatalities to members of the public (including suicides on railways)
- 14,362 non-fatal injuries to members of the public.

16.3 If an accident is reported to the Enforcing Authority, will an investigation take place?

Not always. It depends on the severity of the accident and the approach of the enforcing authorities.

Any accident which involves a major injury is highly likely to be investigated as it shows to the Authority that something serious may have gone wrong with the employer's safety management system.

Sometimes, an enforcing authority will make a telephone investigation first and request further details of management systems so that they can assess your general attitude and commitment to health and safety. If they find information inadequate they will made a site visit.

16.4 What are the consequences if I ignore the law on reporting accidents?

A failure to notify accidents, diseases and dangerous occurrences is an offence under the Regulations and the number of prosecutions for non-compliance is rising. Fines are up to £5,000 in the Magistrates Court.

16.5 What are the type of accidents which have to be notified?

It is not actually accidents which have to be notified, but the consequences of those accidents and the type of injuries which they cause.

Accidents and incidents which arise out of or in connection with work and which fall into the category of:

- Fatality
- Major injury
- Over 3 day injury

must be reported.

Also, any accident or incident which involves a member of the public or non-employee, being sent to hospital also needs to be reported. This is so that information can be gathered on how safe work practices are for members of the public using premises, etc.

Certain types of "dangerous occurrence" must also be reported. These would be incidents which have the potential to cause major or multiple injuries and which could affect large numbers of people – i.e. high risk catastrophes.

Industrial diseases must also be reported within 12 months of the disease being identified.

16.6 What are major injuries?

A major injury must be reported to the Enforcing Authority. Any accident at work or caused by the work activity which results in the following is notifiable:

- Any fracture of a bone (other than finger, thumb or toes).
- Any amputation.
- Dislocation of the shoulder, hip or knee or spine.
- Loss of sight (whether temporary or permanent).
- A chemical or hot metal burn to the eye or any penetrating injury to the eye
- Any injury resulting from an electric shock or electrical burn (including one caused by arcing) leading to unconsciousness or requiring resuscitation or admittance to hospital for more than 24 hours.
- Injuries leading to hypothermia, heat induced illness or unconsciousness, or requiring resuscitation or requiring admittance to hospital for more than 24 hours.
- Loss of consciousness caused by asphyxia or by exposure to a harmful substance or biological agent.
- Acute illness requiring medical attention or loss of consciousness resulting from the absorption of any substance by inhalation, ingestion or through the skin.
- Acute illness requiring medical treatment where it may be caused by exposure to a biological agent, its toxins or infected material.

16.7 What are "over three day" injuries?

Where a person at work is incapacitated for more than three consecutive days from their normal work owing to an injury resulting from an accident at work, then the accident must be reported.

The day the accident happens does not count in calculating the 3 days. But any days which would not be normal working days e.g. shift days, days off, holiday, weekends, do count in the 3 days.

If an employee remains at work but cannot carry out their usual work i.e. are put on "light duties" then the accident must still be notified.

16.8 Does it matter when accidents are reported or are there strict timescales?

As you would expect, there are strict timescales for the reporting of accidents:

(1) Fatalities – immediately or as soon as possible after they happen.
(2) Major injuries – immediately or as soon as possible after they happen.
(3) Over 3 day injuries – within 10 days of them happening.
(4) Accidents to people who are not at work – immediately or as soon as possible after they happen.
(5) Dangerous occurrences – immediately or as soon as possible after they happen.
(6) Diseases – as soon as apparent and without undue delay.

Any accident, disease or dangerous occurrence which is notified immediately – usually by telephone, Email or fax – must be confirmed in writing on the appropriate form within 10 days.

16.9 Where and to whom should accidents be notified?

Overall, the number of notifications of accidents is generally low and there is serious under-reporting (hence the trend to prosecute for non-compliance).

In order to address this, the HSE have made the reporting of accidents much easier and have provided a "one stop shop" for all employers to report accidents, etc. irrespective of whether their enforcing authority is the HSE or the local authority.

All notifiable accidents, dangerous occurrences and diseases can be notified to the Incident Contact Centre (ICC) on:

Telephone   0845 300 9923 (Office hours only Monday–Friday)
Fax   0845 300 9924
Internet   http://www.riddor.gov.uk
Post   Incident Contact Centre
           Caerphilly Business Park
           Caerphilly
           CF83 3GG

The ICC will forward the details of the accident to the appropriate authority. The ICC will send out confirmation copies of any notifications made.

Common diseases include:

- Mesotheiomia
- Asbestosis
- Industrial deafness
- Carpal tunnel syndrome
- Leptospirosis (Weil's Disease)
- Legionellosis
- Anthrax
- Brucellosis
- Hand arm vibration syndrome (vibration white finger)
- Silicosis
- Bladder cancer
- Pneumoconiosis
- Skin cancer.

16.15 What are dangerous occurrences?

The list of dangerous occurrences is quite long and is included in a Schedule to the RIDDOR.

An indication of dangerous occurrences:

- Collapse, over turning or failure of load bearing parts of lifts and lifting equipment
- Accidental release of biological agent likely to cause severe human illness
- Accidental release of a substance which may damage health
- Explosion, collapse or bursting of a vessel or associated pipework
- An electrical short circuit or overload causing fire or explosion
- An explosion or fire causing suspension of normal work for over 24 hours
- The collapse of scaffolding.

Any dangerous occurrence which has the potential to cause significant harm must be checked to see whether it is notifiable.

It is always better to err on the side of caution and report than not.

16.16 Does any injury which happens to a visitor to site or member of the public have to be reported?

No. Only those accidents and resultant injuries which require the public to be taken to hospital as a result of the accident, and which were caused by the employer's *work activity*.

Injuries which happen to the public or others which are due to carelessness or from something over which they have control will not be notifiable.

The injury must result from an accident "arising out of or in connection with work".

Types of incident which would not be reportable if they caused injury to a person:

- Acts of violence causing injury between fellow workers over a personal argument
- Person dying of a heart attack on the premises
- Visitor tripped over their own bag or luggage
- Acts of violence between customers and visitors.

16.17 Is there a legal duty to investigate accidents?

No, not at present although it is good practice to do so.

Accident reporting is only one part of the process of health and safety management. When an accident or incident occurs it is necessary to fine out what caused it, what went wrong, why and what can be done to ensure that it does not happen again.

There is an implied requirement and duty to investigate accidents because Risk Assessments have to be reviewed regularly and when circumstances change. An accident may be "changed circumstances".

Also, with the increase in civil claims, Insurance Companies are forcing employers to investigate the causes of accidents so that strategies can be put in place to prevent future occurrences. This will help bring down Employers Liability Insurance premiums, or at the very least, prevent then rising astronomically.

16.17.1 What are the key steps in an accident investigation?

Every employer should have an accident investigation plan as part of their health and safety management policy.

Accident investigation should be looked upon as identifying what happened and why so that a re-occurrence can be prevented:

- *Step 1*: define the purpose of the investigation
- *Step 2*: define the procedure
- *Step 3*: define what equipment will be needed
- *Step 4*: define how the investigation is to be carried out and what information will need to be gathered
- *Step 5*: define the content of the report
- *Step 6*: decide how recommendations will be implemented.

It is sensible to create an "accident investigation kit" so that everything you need is in one place and valuable time is not lost in trying to find equipment.

An Accident Emergency Investigation Kit

Contents:

- Report form
- Routine checklist for basic questions or prompts to the investigator
- Notebook, pad, paper, pen
- Tape recorder for on-site comments or to assist at interviews
- Camera – instant or digital to take immediate photos of the scene of the accident
- Measuring tape e.g. surveyors tape, builders tape
- Any special equipment regarding the work environment which could assist the investigation e.g. noise meters, air sampling kits
- Witness forms for statements.

16.17.2 What issues will I need to consider when an accident investigation is carried out?

If there is an accident investigation procedure in place you should follow the specific guidelines.

If not, some pointers are:

- Where the accident happened – describe exactly
- Who was injured – were they employee, contractor, public
- What were they doing
- What equipment were they using
- What time was it
- What were the environmental conditions
- What was the condition of the area or equipment
- Are any defects noted e.g.:
  - Maintenance issues
  - Worn flooring/trip hazards
  - Poor lighting
  - Broken guarding
- Was there a safe system of work in place
- Where Risk Assessments available
- Had COSHH (Control of Substances Hazardous to Health Regulations 2002) Assessments been completed
- Who witnessed what happened
- What did they see happen, or what did they hear
- What training had the employee had
- What actions were taken immediately after the incident
- Was the accident notifiable
- What needs to be done to prevent it happening again

- Had equipment been routinely checked
- Are maintenance records available
- Had the work process been regularly reviewed and checked as part of safety monitoring
- Did someone not do something they should have
- Were contractors involved. Had they had induction training and been made aware of any site specific hazards.

17

Site Welfare Facilities

17.1 Is there a legal duty to provide welfare facilities on a construction site?

Yes. The CDM Regulations 2007 contain specific requirements for the provision of welfare facilities on all construction sites.

Generally, everyone who works on a construction site must have access to:

- Sanitary accommodation
- Washing facilities
- Warming facilities
- Somewhere to eat their food
- Somewhere to store clothes
- Drinking water.

The Regulations tend to be "goal setting" which means they do not stipulate exact numbers of facilities for every site. "Suitable and sufficient" is a term which is often used and the employer has to determine what this might be.

17.2 How is "suitable and sufficient" or "reasonable" determined?

To an extent, common sense should prevail.

There must be enough facilities for everyone to use them without excessive waiting, etc.

Guidance on appropriate numbers is provided by the Health and Safety Executive (HSE) or within other documents e.g. British Standards. Employers are expected to know about the existence of other guidance and would be expected to consult good practice guides.

Should a Construction Inspector from the HSE visit site and declare that facilities are inadequate or not suitable and sufficient, an Improvement Notice could be served requiring the provision of additional facilities.

If the employer feels that the Inspector is being unreasonable, he could appeal the Improvement Notice and the matter would be heard at a Tribunal. Suitable and sufficient could be determined in this arena, or ultimately, in a Court of Law.

17.3 Do facilities have to be provided on the site or could toilets available in the area be used?

In the majority of instances, it will be expected to provide welfare facilities actually on the construction site. It will be unreasonable to expect operatives to walk off site to find public conveniences, etc.

However, it may be reasonable to make proper arrangements with another employer to use their facilities elsewhere in the building for instance.

The HSE Inspector would expect to see a proper arrangement covered in the Construction Phase Health and Safety Plan (if the CDM Regulations apply to the project) or within information given to employees.

17.4 How many water closets, urinals and wash hand basins have to be provided?

The CDM Regulations 2007 state the facilities should be "suitable and sufficient".

Guidance on actual numbers of facilities can be found in other documents, namely the Approved Code of Practice for the Workplace (Health, Safety and Welfare) Regulations 1992 and British Standard 6465.

A good starting point for calculating the number of facilities is to use the following tables:

No. of men at work	No. of water closets	No. of urinals	No. of wash stations
1–15	1	1	2
16–30	2	1	3
31–45	2	2	4
46–60	3	2	5
61–75	3	3	6
76–90	4	3	7
91–100	4	4	8
Above 100	An additional WC for every 50 (or part) men plus an equal number of additional urinals, plus an additional wash hand station for every 20 operatives.		
	Wash hand stations should be provided with adequate supplies of hot and cold running water. Water closets should preferably be wash down water types.		

17.5 Do toilets always have to be plumbed in with running water?

No, not always but it is preferable to have flushing toilets if at all possible.

If water supply and drainage cannot be provided to the site welfare facilities it will be acceptable to provide chemical closets.

Suitable numbers of chemical closets must be provided and they must have suitable mechanisms for maintaining the closets in a sanitary condition.

The Site Agent must ensure that greater attention is given to ensuring that chemical closets are kept clean and regular cleaning schedules must be in place. Regular emptying of the sewage containers will be necessary and plans will need to be made to deal with this.

If drains are available for discharging chemical closets the question might be asked as to why flushable toilets cannot be used.

17.6 What hand washing facilities would be acceptable if no running water is available?

Suitable facilities for washing hands and arms are essential on a construction site because the risk of hand to mouth infection is high.

If no running water is available, suitable containers of water must be provided e.g. plastic containers with tap usually associated with camping and caravanning.

If hot water can not be provided it is important to provide anti-bacterial soaps which work in cold waters, or water-less hand gel which effectively sanitises the hands.

Cement dust particularly needs to be removed from hands and arms so as to prevent skin diseases.

17.7 Do urinals have to be provided?

Not necessarily, as the overall number and accessibility of facilities is the most important factor. Urinals can be provided in addition to water closets and where they are, a slightly lesser number of water closets will be needed.

The following table could be used to aid calculations.

No. of men at work	No. of water closets	No. of urinals	No. of wash stations
1–15	1	1	2
16–30	2	1	3
31–45	2	2	4

(Continued)

No. of men at work	No. of water closets	No. of urinals	No. of wash stations
46–60	3	2	5
61–75	3	3	6
76–90	4	3	7
91–100	4	4	8
Above 100	An additional WC for every 50 (or part) men plus an equal number of additional urinals, plus an additional wash hand station for every 20 operatives. Wash hand stations should be provided with adequate supplies of hot and cold running water. Water closets should preferably be wash down water types.		

Source: British Standard 6465 : Part 1 : 1994 Section 7.6 Table 4.

17.8 Whose duty is it to calculate the number of sanitary facilities – the Client or the Principal Contractor?

All construction projects fall under the CDM Regulations 2007 and the Client may stipulate what they expect to see in respect of site facilities. Clients are expected to set the standards for improving overall conditions on construction sites and under CDM 2007 they must approve the provision of welfare facilities before a project starts on site.

If a Client has set the standards these will be found either in the Pre-Construction Health and Safety Pack or in accompanying "Employers Requirements".

The requirements listed in the Pre-Construction Health and Safety Pack should be taken forward by the Principal Contractor and developed into the Construction Phase Health and Safety Plan.

If a Client has not stipulated any specific requirements, it will be assumed that the Principal Contractor will be responsible for ensuring that legal requirements in respect of facilities on the site are met.

The provision of site welfare facilities must be clearly covered in the Construction Phase Heath and Safety Plan. If the Client feels that insufficient provision is made he can prohibit start on site of the works under Regulation 16 CDM as the Construction Phase Health and Safety Plan will not be considered sufficient.

The Client should give an indication to the Principal Contractor of the anticipated number of contractors or workers expected on the site especially if any of these will be client direct appointments or nominated contractors. This should help the Contractor to calculate numbers of facilities.

17.9 If site welfare facilities are shared between all contractors, who is responsible for keeping them clean?

The Construction Phase Health and Safety Plan should detail who is responsible for providing and maintaining welfare facilities. The bigger and more complex the project the greater the requirement for detailed information on welfare facilities to be provided.

Generally this will be the Principal Contractor although another contractor could be identified as being responsible.

On notifiable CDM projects, the Principal Contractor is responsible for ensuring the co-ordination and co-operation of employers on a multi-occupied site and there must be a clear indication of who is responsible for providing and maintaining site welfare facilities.

Where facilities are shared with a residual employer e.g. refurbishment projects in occupied buildings, agreement must be reached between both sides regarding who is responsible for cleaning and maintenance, or indeed, whether the facilities can be shared. Sometimes an influx of site workers will cause existing facilities to become inadequate and the Principal Contractor may have to provide additional facilities.

If facilities are found to be in unsatisfactory condition, not clean, etc. and an HSE Inspector visits site, it is likely that an Improvement Notice will be served under the Health and Safety at Work Etc Act 1974.

17.10 Is it necessary to provide separate sanitary and washing facilities for women?

Men and women may use the same toilet provided it is in a separate room with a door which can be locked from the inside. Where possible, cubicle walls and door should be full height, floor to ceiling so that the cubicle is totally enclosed.

Wash hand basins can be shared between the sexes for hand and arm washing. It would be good practice to have a water closet and wash hand basin in one cubicle but if this is not practicable, communal wash hand basins in an anteroom to the water closets would suffice.

17.11 What provision needs to be made for Clients and visitors?

There is no legal requirement for separate facilities to be provided for Clients and visitors although many sites do have separate facilities.

A Client may stipulate that separate facilities are required and this will be a matter of agreement between contractor and Client. The HSE Inspector will only be concerned with the number of welfare facilities on site for the number of operatives working on the site.

Often separate facilities are provided because they can be locked shut and kept in a more acceptable condition.

17.12 What provision needs to be made on a construction site for drinking water?

Schedule 2 of the CDM Regulations 2007 requires that a suitable supply of drinking water be provided on every construction site.

Drinking water is "potable" water and meets the requirements of the Drinking Water Regulations. It generally needs to be a mains piped supply but adequate quantities of bottled water or water containers and dispensers will be satisfactory.

Every supply of drinking water shall be conspicuously marked by an appropriate sign where necessary for health and safety reasons. This is especially important where there may be two or more supplies of water around the site and one is fit for drinking whilst the others are not.

Where a supply of drinking water is provided there must also be provided a sufficient number of cups or other drinking vessels unless the water supply is from a purposely designed drinking fountain.

On large sites, a suitable supply of drinking water must be provided at readily accessible and suitable places. This could be within each floor if the construction site is a multi-storey building, or within say, every 100–200 metres.

It would *not* be considered reasonably accessible for only one drinking water supply to be available in the canteen if several floors need to be climbed to get to the facility.

More drinking water points need to be provided in warmer weather than in winter. If the site has a "no food and no drink on site" rule it is imperative that the Principal Contractor makes adequate provision for drinking water supplies.

17.13 What facilities are required for the changing and keeping of clothing on site?

Schedule 2 of the CDM Regulations 2007 requires that "suitable and sufficient" changing rooms be provided or made available:

(a) For clothing of any person at work on the construction site and which is not worn during working hours and

17.16 What provision for heating has to be made of a construction site?

Indoor working temperatures have to be reasonable – there is no maximum or minimum temperature set down in the CDM Regulations 2007 (nor indeed in other Health and Safety legislation although there is guidance).

Reasonable working temperatures are a matter of interpretation of a number of circumstances:

(i) What work needs to be done – how physical is it?
(ii) How open is the site to the elements?
(iii) How many people on site?
(iv) How easy is it for operatives to leave the work area to get warm?
(v) How much thermal clothing can be worn comfortably?

Generally, in order to arrive at a suitable decision as to what level of heating is required, a risk assessment will be necessary.

The Principal Contractor should complete the risk assessment for all aspects of the site and must communicate the findings to all contractors.

The hazards from too cold a working environment are:

(i) Increased risk of accidents
(ii) Lack of concentration
(iii) Increased risk of heart attack
(iv) Hypothermia
(v) Poor circulation of the blood.

Obviously, the biggest concern is the likely increase in accidents if people are too cold to hold tools effectively, mix materials, etc.

The Principal Contractor must identify what control measures can be put in place to reduce the hazards and risks associated with low working temperatures.

Space heating could be installed in suitable locations throughout the site. Heating appliances are:

(i) Electric
(ii) LPG
(iii) Gas.

Inadequately ventilated LPG and gas heaters could cause carbon monoxide gas to be produced and this could lead to fatalities. Gas equipment may continually leak because valves have been left on. These potential hazards need to be weighed up when choosing a suitable heating source.

Electrical fan blowers purposely designed for construction sites and running on 110 volts may be the safest option.

17.16.1 Tragic incident: Carbon monoxide poisoning

In December 2006, three men died from carbon monoxide poisoning on a construction site. The men were using a steel container as their rest room and were running a generator to provide the heating to keep warm. The steel container had no ventilation and the three men were sleeping on site. Carbon monoxide was produced due to the incomplete combustion in the generator and dangerous levels of carbon monoxide can accumulate within minutes, especially in a confined space.

All construction site operators must be aware of the risks. All flueless open flame heaters fuelled by natural gas or LPG require an adequate supply of fresh air to prevent the formation of high levels of carbon monoxide.

17.16.2 What provision for ventilation has to be made for a construction site?

All workplaces must be provided with adequate ventilation which means a supply of purified air or fresh air. Construction sites are no exception although generally, there is little difficulty in providing adequate ventilation due to the open nature of many sites.

Particular attention has to be paid to the adequacy of ventilation when dust and chemicals, fumes and vapours are produced around the site.

Extract ventilation may be needed at certain times to eliminate dust or fumes. This requirement should be covered in the COSHH (Control of Substances Hazardous to Health Regulations 2002) Assessment completed for the work activity which would generate the dust or fumes.

It would be sensible to complete a risk assessment to determine the needs for ventilation. Certain areas of the site may be "confined spaces" and additional provision may be required.

Increased ventilation may be needed in summer months – a constant review of conditions on site is necessary so that adaptations to ventilation requirements can be made.

17.17 What provision needs to be made for lighting on the site?

Regulation 44 CDM Regulations 2007 sets out the requirements for lighting on a construction site.

As expected, lighting on site needs to be "suitable and sufficient". Lighting should also be, where practicable, natural light.

If artificial lighting is used it must not cause any warning signs or symbols to be adversely affected by a change of colour, etc.

Lighting must be suitable and sufficient and the only test is "can you see where you are going around site?"

Is it possible to see the floor clearly, any small holes, drainage channels, etc. Can the task at hand be seen clearly without eye strain? Can operatives see what they are doing? Can they see the emergency exit signage? Can they see clearly when using steps and stairs or are they at risk of falling? Is there a difference between outdoor sunlight and internal lighting which may cause temporary blindness?

Is artificial lighting in the right place? Does it have enough brightness or lux?

Poor lighting increases the risk of accidents. Improve the lighting levels and the accident rate may reduce causing the site to be much more efficient.

Early morning starts and dusk in winter need to be considered and more lighting will be necessary. Do not forget to consider entering a darker building from external sunlight – eyes need time to adjust so make sure that there is more than enough lighting.

Is it clear what lighting is to be provided by whom? It should be clearly laid out in the Construction Phase Health and Safety Plan. The Principal Contractor is responsible for background lighting and for lighting means of access and egress to the site. Contractors and sub-contractors may be responsible for providing task lighting for their individual trades.

The Principal Contractor should discuss the aspect of lighting in pre-start meetings and should check the contractor's risk assessments to make sure that they have considered the need for task or background lighting.

The Principal Contractor will also be responsible for ensuring that adequate provision is made on the site for secondary lighting when it is considered there would be a risk to health and safety if the primary lighting failed.

Emergency lighting will be required on fire exit routes. The Fire Plan should deal with these issues.

18

Asbestos

18.1 What are the main duties under the Control of Asbestos Regulations 2006?

Employers have duties under the Regulations to protect their employees from exposure to asbestos-containing materials as they may cause harm to health.

Under the Regulations, employers are responsible for the health and safety of:

- Their employees
- Other peoples' employees
- Members of the public
- The self–employed.

If they are or will be exposed to asbestos.

Employers also must:

- Provide information, instruction and training
- Carry out risk assessments
- Produce a written plan of work
- Ensure asbestos types are identified
- Prevent or reduce exposure to asbestos
- Introduce control measures
- Maintain effectively and control measures
- Keep records of any tests, examination
- Provide suitable protective clothing
- Provide changing facilities and clean clothing
- Develop emergency procedures
- Prevent or reduce the spread of asbestos
- Clean equipment and premises after exposure to asbestos
- Designate areas "respirator zones" or an "asbestos area"
- Display suitable hazard warning notices
- Arrange for effective air monitoring
- Keep records for 5 years or for 40 years if to do with health surveillance
- Provide health surveillance every 2 years to those exposed to asbestos
- Remove asbestos waste under special waste provision.

Contravention of any of the Regulations is an offence and fines can be up to £5,000 per offence in the Magistrates Court, or for serious offences and

A duty holder is defined in the Regulations as:

- Every person who has, by virtue of a contract or tenancy, an obligation of any extent in relation to the maintenance or repair of non-domestic premises, or any means of access thereto or egress there from; or
- In relation to any part of non-domestic premises where there is no such contact or tenancy, every person who has, to any extent, control of that part of those non-domestic premises or any means of access thereto or egress there from.

Where there is more than one duty holder, the relative contribution to be made by each person in complying with the requirements of the Regulation will be determined by the nature and extent of the maintenance and repair obligation owed by that person.

A wide range of people will potentially have obligation under the Regulation, including:

- Employers
- Self-employed
- Owner of premises
- Managing agents.

The duty to manage asbestos does not extend to domestic premises but does apply to residential premises if they are let as a business e.g. hotels, bed and breakfast establishments, caravan parks and so on.

18.7 As a duty holder what do I need to do?

Find out if asbestos-containing materials are present in your building or premises.

If the building was constructed before 1985 it is likely to contain some asbestos unless it has already been removed.

Buildings constructed up to 1999 may have asbestos cement materials.

18.8 What are the steps to be taken to establish whether asbestos-containing materials are present in the building?

Checking building plans and other information such as operating manuals, etc. to see if any reference has been made to asbestos or asbestos materials.

Consult the Design Team, if possible, who undertook the building works, including any known contractors or sub-contractors, building services contractors, etc.

Carry out a full survey of the premises to identify likely asbestos-containing materials.

18.9 What types of surveys are there which can be used to identify asbestos-containing materials?

It is important to note that the CAW 2006 do not legally require surveys in order for a duty holder to comply with their duties under the Regulations.

But, as the Regulations require asbestos-containing materials to be managed you will have to know where they are, so a survey will be vital.

If you did not want to conduct a survey you could make a "presumption" that all suspect or likely materials will be considered to contain asbestos and act accordingly. But this could cause you to undertake costly preventative measures when lesser ones may be all that is required.

There are three types of survey usually undertaken to identify asbestos-containing materials in buildings. Whichever survey is chosen, you must always ensure that it is carried out by a competent person. Check their qualifications, accreditations and experiences. Seek references from others.

Some survey reports are riddled with so many exclusion clauses that they become pretty meaningless. Asbestos identification may be a case where "cheapest isn't always best".

18.9.1 Type 1: Location and assessment or presumptive survey

The purpose of this survey is to locate as far as is reasonably practicable, the presence and extent of asbestos-containing materials, including their condition.

Samples and analysis of materials is generally not undertaken until such time as more detailed information is required.

All areas should be accessed and inspected.

If it is unclear as to whether material contains asbestos it is to be assumed that it does until proved otherwise.

18.9.2 Type 2: Standard sampling, identification and assessment or sampling survey

Under this survey type, asbestos-containing materials are positively identified through sampling and analysis. Visual inspection is carried out as for Type 1, but samples from each type of asbestos-containing material are taken by the surveyor.

Results of the analysis will identify exactly what is and is not asbestos material. A management plan can then be actioned.

18.9.3 Type 3: Full access, sampling and identification or pre-demolition/major refurbishment survey

This survey is undertaken before all major building works as it requires that an extensive survey of the building is undertaken, with full destructive inspection

Do not disturb and break into material to see if it contains asbestos – you often cannot tell by looking at it and if it is asbestos you expose yourself and others to health risks from fibres.

If your initial survey indicates that materials are likely to contain asbestos then you will need to have a full destructive survey carried out whereby samples of the material are taken for analysis. Only trained and competent persons should undertake sampling.

18.13 What procedures need to be taken if asbestos-containing materials are found?

Assess its condition by:

- Is the surface of the material damaged, frayed or scratched?
- Is any part of the material peeling or breaking off?
- Is it detached or loose from the structure it was applied to e.g. falling away from pipes, structural steel?
- Are protective coatings and coverings damaged?
- Is asbestos debris or dust evident in the area?

18.14 What steps need to be taken if asbestos material is in poor condition?

Asbestos in poor condition is a serious health hazard. You must either:

- Remove it
- Repair it
- Encapsulate it
- Seal it.

Licensed contractors will need to be appointed to work on most asbestos materials. You must agree with them an Action Plan and must ensure that all legal requirements are met. Any asbestos-containing material in poor condition is best dealt with by a competent, licensed contractor.

18.15 What steps need to be taken where asbestos-containing materials remain in the building?

An Asbestos Management Register must be created to help ensure all asbestos is located and regularly inspected.

Display "Hazard Warning Signs" on all residual asbestos or some other form of identification.

Operate Permit to Work procedures for maintenance works, refurbishment works, etc.

It is not illegal to have asbestos-containing materials on the premises, provided they do not cause a health hazard. It is safer to have asbestos in good condition in situ and manage the potential risk, than to remove it and create high-health risks.

18.16 What are some of the steps for managing asbestos?

Make regular checks of the condition of the asbestos material to ensure that it is not deteriorating.

Carry out Risk Assessments for any work in the area e.g. could anything puncture the asbestos material releasing fibres and so on?

Keep records up-to-date and show that an Asbestos Management Plan is in place.

Introduce a Permit to Work system for all contractors, maintenance engineers, etc. so that you know where they will be working, why, on what and what the hazards and risks are. If they are to work near asbestos material this can be highlighted and safety precautions stipulated.

18.17 How often do I need to check the condition of asbestos material?

Once every 12 months will usually be sufficient, but if the asbestos is in the area which is at risk of damage, then more frequent inspections will be required.

18.18 How do I dispose of asbestos waste?

Asbestos comes under the Special Waste Regulations 1996 and can only be removed to a licensed waste disposal site.

Asbestos must be double bagged and sealed in heavy duty polythene bags and clearly labelled with a recognized asbestos label.

Case Study

HSE Inspectors successfully prosecuted two companies under the CDM Regulations 1994 for failing to manage asbestos on site. Fines were more than £20,000.

The Client had failed to conduct a proper survey of a boiler house which contained asbestos and in which work was to be carried out. Had a suitable survey been carried out it would have revealed the presence of asbestos. The Client was prosecuted and fined £10,000 plus nearly £5,000 in costs.

The contractor was also prosecuted for putting its workers at risk by exposing them to asbestos.

Neither company had followed proper procedures. The Client had relied on third-hand information from another contractor which included analysis results from unrepresentative samples. Pipework was dismantled by operatives and put in black sacks. No-one had any personal protective equipment. There were no safe systems of work nor any competent person overseeing the work.

Asbestos was discovered when a licensed asbestos contractor was called to site and air samples revealed large amounts of asbestos fibres in the air.

The HSE commented that the Client had a legal duty to carry out a proper survey, relay information to the contractor and employ competent persons to remove the risk.

19

Safe Places of Work

19.1 What are the duties of the Site Agent in respect of safe means of access to site?

Under Section 2 of the Health and Safety at Work Etc Act 1974, an employer is responsible for ensuring that employees and others have safe means of access and egress to their place of work.

Regulation 26 CDM 2007 also requires safe access and egress to every place of work.

A "place of work" can be anywhere where an individual is expected to perform their duties and can include buildings, rooms, open spaces, working platforms, roofs, scaffolding, etc.

The Site Agent has a duty on behalf of his employers to ensure that employees have safe means of access and egress. In addition, the Site Agent has to ensure that persons other than those in his employ have safe means of access and egress.

Safe means of access mean access without risk of injury or harm.

There should be no risk of being run over by vehicles, no risk of tripping or falling over materials, plant, etc., no risk of falling from any height, no risk of objects falling on to people using the access route.

Safe access needs to be considered in relation to whether tools and materials need to be carried to the place of work. Vehicles colliding with pedestrians is a common accident cause on construction sites.

Under the CDM Regulations 2007 the Principal Contractor has the duty to prevent unauthorized access to site.

The CDM Regulations apply to all construction projects and the Site Agent must include in the Construction Phase Health and Safety Plan the details of safe access to site e.g. the preferred pedestrian route, how and where materials and plant will be delivered, vehicular access, routes out of site, emergency exits and so on.

19.2 What is the requirement for safe access and egress under CDM 2007?

Regulation 26 CDM 2007 states that:

"There shall, so far as is reasonably practicable, be suitable and sufficient safe access and egress from every place of work and to and from every other place provided for the use of any person while at work, which access and egress shall be properly maintained.

Suitable and sufficient steps shall be taken to ensure, so far as is reasonably practicable, that no person uses access or egress, or gains access to any place, which does not comply with the requirements of paragraph (1) or (2) respectively".

Place of work is defined as any place which is used by any person at work for the purposes of construction work or for the purposes of any activity arising out of or in connection with construction work.

Places of work can therefore be:

- The whole construction site
- Specific areas of the site
- A work platform
- A roof
- A ladder
- A crane
- An elevating platform
- A trench or excavation
- A confined space
- The mess room
- The site office.

19.3 Can pedestrians and vehicles use the same access routes?

It is important to keep vehicles and pedestrians as separate as possible as a high number of accidents, including fatalities, occur when vehicles and pedestrians share access routes.

Regulation 36 of the Construction (Design and Management) Regulations 2007 requires that construction work be so organized that, where practicable pedestrians and vehicles can move safely and without risks to health.

In particular, traffic routes will not be considered safe if:

- Steps have not been taken to ensure that pedestrians, when using traffic routes, can do so without danger to their health and safety.

- Any door or gate opens directly into the traffic route without any clear view of approaching traffic or vehicles.
- Pedestrians cannot have a place of safety to view oncoming vehicles.
- There is no adequate separation between pedestrians and vehicles.

So, pedestrians and vehicles can use the same traffic route but *strict* safety rules apply and generally, a Risk Assessment will be necessary which identifies the hazards and risks.

19.4 What precautions can be taken to separate pedestrians and vehicles, or manage the risks to their health and safety?

A number of safety precautions can be taken to ensure the safety of pedestrians:

- Designate a safe walking area by painting hatch markings on a coloured walkway.
- Put up guard rails to delineate the walkway.
- Use a banksman to direct vehicles and pedestrians.
- Use mirrors so that drivers and pedestrians can see routes clearly.
- Ensure all vehicles have audible warning devices – not only when they are reversing.
- Have a policy for all vehicles to ensure that lights and hazard lights are on during all hours, not just dawn and dusk.

19.5 What will constitute a safe means of access to upper work levels?

Where practicable, permanent means of access to upper levels will be expected e.g. installation of the permanent staircase is preferable.

If the permanent structure cannot be installed than a purposely designed temporary staircase is preferable.

When determining safe means of access, remember to consider what people have to carry to their place of work e.g. materials and tools.

Ladders are not necessarily classed as a safe means of access, especially for longer term projects. Ladders give immediate access to higher levels – they are not necessarily safe. Please refer to Chapter 24: Working at Height.

Safe means of access to upper levels may be by way of hoists and lifts.

Mobile elevating platforms are safer for accessing high-level works e.g. ductwork, ceiling works, etc. They also provide a safe working platform.

19.6 What other precautions need to be considered for safe access and egress to the place of work?

Access routes must:

- Be clearly lit at all times
- Be free of obstructions and trailing cables
- Have clearly defined steps or slopes
- Have handrails if significant changes in level
- Be of adequate size for the number of operatives using them
- Be protected form hazards
- Not to be underneath activities being carried out at height e.g. under a crane sweep and so on
- On stable ground
- Clearly visible
- Designed to be away from material delivery points, etc.
- Identified in the Pre-Construction Health and Safety Pack
- Consider any other access routes into the building e.g. if the employer is still operating out of the building
- Not create crush points during clocking on/off times
- Be adequately signed so that people know where they are expected to go
- Have designated crossing points if vehicle/traffic routes need to be crossed
- Be highlighted on a plan at the entrance to the site.

19.7 Is it necessary to provide a security point and signing in station to a construction site in order to manage safe access?

It is a good practice to have a site control point because it is necessary to know who is and is not on site at any one time for fire safety purposes.

As a minimum, a designated signing in place is essential at the site entrance. A site log for the signing in and out of site is essential. This log can act as a reference log for fire safety purposes and will help to check whether everyone is accounted for in the event of a fire.

On complex sites, it is a good practice to have a full-time security person operate the control point. This allows not only strict control of persons on to and off site, but also allows for checks on deliveries, skip removals, etc. and may help the issue of materials and equipment thefts.

The Pre-Construction Health and Safety Pack should indicate whether the Client expects the Principal Contractor to provide a full-time operated access control point.

The Principal Contractor has to take responsibility under the Construction (Design and Management) Regulations 2007 for preventing the unauthorized persons access to site. A properly controlled access control point will discharge this responsibility and even if someone should gain access to site and put themselves in danger, the Principal Contractor would have a defence.

There has been a great deal of development over the years in computerized security systems. These are based on a swipe card and computer database and record access times and exit times. Many systems also combine photo identity cards and are used for training record purposes.

19.8 Does the construction site have to have more than one exit point?

The number of exit routes from a construction site depends on the number of operatives working on the site and the size and complexity of the site.

The Construction (Design and Management) Regulations 2007 stipulate that a suitable and sufficient number of exit routes must be provided which can be used in an emergency. Any person on the site must be able to reach a place of safety quickly and without hindrance.

It is always a good practice to have more than one exit route. The access or way in can also double up as the exit route and often, this will be the most popular route, as it is well known. There should be alternative exits form places of work if the travel distance to an exit route is more than 45 metres – if a clear run, or more than 30 metres if the route is less direct.

On large sites, several exits from the place of work will be needed.

An exit route cannot be counted if it leads people back into the building or site or leads them to a dead end. An exit route must lead to a place of safety.

19.9 Do emergency exit routes have to have emergency lighting?

If emergency exit routes are to be used in poor daylight conditions e.g. dawn and dusk, or daylight is generally poor (as in winter) emergency lighting will be essential on construction site emergency exit routes. Emergency lighting should come on if any artificial lighting fails or if visibility is low. Such emergency lighting must be automatic.

Emergency lighting is particularly essential on staircases leading out of the building – poor lighting could lead to people falling on the stairs causing potential bottlenecks and crushing hazards.

Emergency signs should also be clearly visible and the emergency fire signs should be illuminated. At the very least, photo-illuminescent signs can be used.

For more information see Chapter 30 on Fire Safety.

20

Vehicles and Transport

20.1 Vehicles and mobile plant are in common use on construction sites. What are the main hazards?

Vehicles, mobile plant and pedestrians do not go well together and the vehicles and mobile plant are the causes of many accidents, many with fatal consequences.

The key hazards when using vehicles or mobile plant are:

- Moving vehicles running over operatives
- Over-turning vehicles or mobile plant
- Reversing vehicles
- Vehicles or plant too close to excavation edges
- Vehicles or plant positioned on unstable ground
- Vehicles or plant coming into contact with overhead power lines
- Vehicles or plant coming into contact with buried services
- Restricted access to site
- Restricted vision of vehicle and plant operatives
- Untrained operatives driving vehicles
- Leaving vehicles unattended, with keys in the ignition
- Over loading and therefore over balancing of mobile plant
- Inadequate operating space when using vehicles or mobile plant.

20.2 How can hazards associated with vehicles and pedestrians be reduced to a manageable level on site?

Good site planning will help reduce hazards on the site. This could start at the design stage of the project and should involve the CDM Co-ordinator, where appointed under the CDM Regulations, so that all Designers on the project are aware of what vehicles and plant may be needed on the site.

Provide safe entry and exit points with adequate turning space and good visibility for drivers.

Ensure that there is good lighting and visibility in areas close to pedestrians. Provide additional "street lights", avoid blind corners and obstructions. Keep a good sight line for drivers of vehicles.

Plan to keep vehicles and pedestrians separate by having different entrances and exits for vehicles and people. Make sure each entrance and exit is properly signed, with clear text or pictogram signs which can be seen from all areas.

Provide separate, barriered walkways for pedestrians.

Where vehicles in particular have to be used in close proximity to pedestrians, provide a banksman.

20.3 What procedures need to be in place for reversing vehicles?

More accidents are caused by reversing vehicles than any others and the safety record of any site can be significantly improved by managing reversing vehicles.

Consider a one-way system for the site. This should be done at the Planning Stage and the CDM Co-ordinator needs to co-ordinate with the Designers to consider whether any alterations to the site layout so that a one-way system can be accommodated. It may be necessary to slightly re-locate the position of a building so that adequate access is made available.

If vehicles need to reverse around the site ensure that they are fitted with audible reversing alarms and lights.

In any highly populated area, a banksman wears a high-visibility jacket and is properly trained in the tasks to be undertaken.

Prevent persons from crossing or moving in the area until the vehicle has completed its manoeuvres.

20.4 What are the good practice guidelines regarding route-ways around the site?

It is sensible to set out clear and signed route-ways across the site.

Avoid blind corners, sharp bends, narrow gaps, places with low-head room.

Avoid step gradients, adverse cambers, shafts and excavations.

Provide a temporary road surface – this not only helps delineate the roadway but also helps to provide a level, stable surface on which vehicles travel.

Introduce a regular inspection and maintenance programme for all of the route-ways. Potholes create hazards which could cause vehicles to over-turn or loads to dislodge. Regular repairs need to be instigated.

Keep vehicles away from temporary structures, especially scaffolds as an accidental knock of a scaffold pole could cause the scaffold to collapse.

Erect speed limit signs e.g. 5 miles per hour maximum speed limit throughout the site. Enforce the rules!

Erect directional signs to the entrance, exit or materials/delivery area.

Reduce the amount of mud transferred around the site and ultimately onto the highway by installing wheel washers at key locations e.g. entrances and exits.

Protect any excavations with barriers which both highlight the excavation guard to prevent vehicles falling into the void.

Do not allow other vehicles to park on route-ways, nor for route-ways to be used for the storage of materials or plant.

20.5 What precautions are needed for vehicles which carry loads?

Make sure that vehicles which carry loads have been designed to carry loads – do not improvise and adapt vehicles.

Loads should be securely attached and any loose materials e.g. bricks, timber and so on must be secured with netting or tarpaulin or similar. Materials blowing off vehicles create major health and safety hazards and falling material can cause serious injuries to site operatives and pedestrians.

Vehicles must not be overloaded as they will often become unstable. Vehicles used for carrying or lifting loads will have a *safe working load* limit – make sure this limit is understood and weights are within the limits. Overloaded vehicles are also difficult to steer and their braking efficiency is impaired.

Operatives who drive vehicles must be over 18 years of age.

All operatives must be properly trained to drive vehicles.

Case Studies

A labourer riding as an unauthorized passenger on a dumper, fell and struck his head on the road. He died of head injuries.

An untrained labourer thought that he could drive a dumper truck to collect some debris. The keys were left in the ignition. He drove it over uneven ground, lost control and turned it over on the edge of an excavation. He was crushed to death as it over-turned.

A poorly maintained dumper over-turned into an excavation causing the driver to be trapped underwater. The braking system had failed due to lack of routine servicing.

20.6 Are there any special precautions in respect of health and safety to be taken when using dumper trucks?

Compact dumpers, the official name for site dumpers, are responsible for approximately one-third of all construction site transport accidents.

The three main causes of accidents involving dumper trucks are:

- Over-turning on slopes and at the edges of excavations, embankments, etc.
- Inadequately maintained braking systems.
- Driver error due to lack of experience and training e.g. failure to apply the parking brake, switch off the engine and remove the keys before leaving the driver's seat.

The way to avoid such accidents is to pay attention to the following:

- Ensure that all dumper trucks in use have "roll over protective structures" (ROPS) and seat restraints.
- Ensure that where there is a risk of drivers being hit by falling objects or materials, dumper trucks are fitted with "falling object protective structures" (FOPS).

Both ROPS and FOPS are legally required on all mobile equipment. If the equipment is hired, the hire company must ensure that the equipment complies with the law.

- Ensure that a safe system of work operates when using dumper trucks.
- Provide method statements and risk assessments for using dumper trucks.
- Ensure that a thorough maintenance check is carried out, and in particular, that the braking system is checked.
- Check drivers' training records, driving licence and general competency to drive vehicles.
- Operate a strict no alcohol or drugs policy and prohibit anyone from driving who appears unfit to do the job.
- Plan any additional precautions necessary for using dumper trucks in inclement weather. Restrict their use in icy conditions.
- Provide adequate stop blocks to prevent dumper trucks falling into excavations, etc. when tipping.

20.7 What are the key requirements for operating forklift trucks safely?

Every year there are about 8,000 accidents involving forklift trucks (HSE Information) and an average, ten of them, involve fatalities.

There are a few simple measures which can be taken to manage the use of forklift trucks on a construction site, namely:

- Manage lift truck operations using safe systems of work
- Provision of adequate training for operators, supervisors and managers
- Using suitable equipment for the job to be done
- Laying out premises in such a way as to ensure that lift trucks can move safely around
- Ensuring that lift trucks are maintained safely
- Ensuring that the premises and site in which they are to be used are maintained in safe conditions e.g. pot holes in access roads infilled and so on.

20.8 What legislation applies to the use of forklift trucks?

The main legislation includes:

- Health and Safety at Work Etc Act 1974
- The Management of Health and Safety at Work Regulations 1999
- The Provision and Use of Work Equipment Regulations 1998
- The Lifting Operations and Lifting Equipment Regulations 1998
- The Workplace (Health, Safety and Welfare) Regulations 1992.

Breaches of sections of the Health and Safety at Work Etc Act 1974 can incur fines of up to £20,000 and for the other Regulations; fines are up to £5,000 per offence.

20.9 How do I know if a lift truck operator is properly trained?

All lift truck drivers and operators must be able to demonstrate that they have received proper training from a competent instructor and from a recognized training body.

Operatives should have a valid certificate of training issued by one of the following:

- Association of Industrial Truck Trainers
- Construction Industry Training Board
- Lantra National Training Organization Ltd
- National Plant Operators Registration Scheme.

Operators should be able to show the three stages of training as being completed:

- Basic
- Specific job
- Familiarization

Basic training should cover the skills and knowledge required to operate a lift truck safely and efficiently.

Job training should be site specific and should include:

- Knowledge of the operating principles and controls of the lift truck to be used especially where these relate to handling attachments specific to the job.
- Knowledge of any differing controls on the machine to be used as this may be different to the one they were trained on.
- Routine servicing and maintenance of the lift truck in accordance with the operator's handbook as may be required to be carried out by the operator e.g. pre-start safety checks, visual checks, oil and brake checks and so on.
- Information on specific site conditions e.g. slopes, overhead cables/beams and so on; excavations, one way vehicle routes, confined spaces, designated exits and so on; speed limits, site rules and so on.
- Details of the tasks to be undertaken, type of loads to be carried, hazardous areas, materials, unloading and unloading areas, etc.

Familiarization training should take place on the site with the operator being supervised by a competent person. A full "walk through" the site is recommended so that hazardous areas, site layout can be explained. Also emergency procedures must be covered during site familiarization.

Even when operators have formal certificated training qualifications in operating forklift trucks it is sensible to record on a site training record the subjects covered on the site familiarization stage and have the operator sign acknowledgement.

Case Study

An employer had to pay out nearly £30,000 in fines and costs for an accident involving the operator of a forklift truck.

The employee who was driving the truck was crushed to death when it over-turned.

A hole had been dug in the yard of a new building being constructed to lay the sewer pipes. On the day of the accident, the employee was using the forklift truck to load goods onto the back of a lorry. He reversed the lift truck across the yard but one of the wheels slipped into the hole and the vehicle toppled over. The driver was trapped beneath the vehicle and suffered severe crushing injuries and died at the scene.

The employer failed to ensure that the surface of the yard was suitable for a forklift truck. Everyone knew about the hole but it was getting bigger due to water erosion. The employer had no system in place to monitor the condition of the yard and had not required a safe system of work, nor had he required Risk Assessments.

Prosecutions were brought under the Workplace (Health, Safety and Welfare) Regulations 1992, Regulation 12.

Top Tips

Safe Driving Practices

- Check tyres, brakes, operating systems on all vehicles at the beginning of every day.
- Wear protective clothing and equipment e.g. ear defenders, high-visibility jackets.
- Use dumper trucks with ROPS and FOPS.
- Use vehicles with reversing alarms and adequate lighting.
- Follow site speed limits.
- Check that any loads are evenly distributed.
- Do not overload vehicles.
- Know the characteristics of the vehicle in all weather conditions.
- Ensure that all drivers are trained and competent.
- Separate pedestrians from vehicles.
- Do not stand on vehicles when they are being loaded or unloaded.
- Select neutral gear, switch off engine and remove keys when stopping and leaving any vehicle.
- Keep to designated route ways.
- Make sure stop blocks are used.
- Do not use vehicles on steep inclines, adverse cambers, etc. without planning the job safely.
- Consider the unexpected and have a plan of action ready.
- Ensure good visibility at all times – do not overload dumpers, etc. so as to restrict vision.
- Check the area via mirrors, shouting, etc. before moving off from an area.
- Do not drive any vehicle if unfit to do so for any reason.

21

Excavations and Demolitions

21.1 What are the legal requirements governing excavations?

Regulation 31 of the CDM Regulations 2007 sets out the requirements for excavations.

Firstly though, Regulation 2 of the above Regulations must be consulted in order to define the term "excavation".

In Regulation 2 (Definitions) an excavation is deemed to include:

- Earthworks
- Trench
- Well
- Shaft
- Tunnel
- Underground working.

Regulation 31 sets out the following:

"All practicable steps shall be taken, where necessary, to prevent danger to any person, including where necessary, the provision of supports or battering to ensure that:

(a) Any excavation or part of an excavation does not collapse accidentally
(b) No material form a side or roof, or adjacent, to any excavation is dislodged or falls
(c) No person is buried or trapped in an excavation by material which is dislodged or falls.

In addition, suitable and sufficient steps shall be taken, where necessary, to prevent any part of an excavation, or ground adjacent to it, from being overloaded by work equipment or material."

The final requirement in relation to excavations is to ensure that all supports and battering are properly inspected by a competent person.

Protect the excavation whilst operatives are working in it so as to prevent persons or equipment falling in and potentially crushing the operatives.

When work in an excavation is finished for the shift it might be safer to full board over the hole so as to prevent any risk of persons falling. Consideration must obviously be given to span widths, etc. so as not to create an even greater hazard from unsafe surfaces.

Keep all people, vehicles, plant and equipment away from excavations. Create an "Exclusion zone" around the excavation. Barrier off the excavation well beyond the edges. Erect hazard warning signs. Make sure that the excavation can be clearly identified. Consider any hazards of poor lighting, restricted access, traffic routes and walkways, etc.

Complete risk assessments for any tipping activity into the excavation and ensure that *stop blocks* are used to prevent vehicles over-running. Consider whether hand balling infill spoil, etc. will be safer and stop the lorry away from the excavation edge.

Remember health and safety is about *accident prevention*, so introduce safety procedures over and above the bare minimum e.g. excavations shallower than 2.0 metre may need edge protection.

21.8 What are some of the other hazards relating to excavations which need to be identified?

Excavations often involve the use of excavators, diggers, etc. and often, more accidents are caused because people are killed by the vehicle or plant than by falling into or being crushed by the excavation.

Keep workers separate from moving plant. Where this is not possible, employ a banksman to guide the excavator and to protect people. Not all operatives on a site will be familiar with the hazards associated with excavations.

Develop safe systems of work which manages when, how, where and by whom excavation work is undertaken and decide how such work may impinge on the safety of others.

Structures often become undermined because of excavation works and adjusting buildings or structures can collapse into the excavation or in close proximity to it.

Structural Engineers should be consulted when excavations are to take place adjacent to buildings or structures. Details of foundation depths, etc. should be given to the Principal Contractor via the Pre-Construction Health and Safety Pack and/or the CDM Co-ordinator.

21.9 What is a safe means of access and egress for excavations?

The bottom of an excavation is often the place of work for someone and a safe means of getting in and out is essential. Accidents happen because access ladders, etc. are not secured or there is no safe way of getting materials and equipment into the trench/excavation.

A ladder is probably the most common method of access into an excavation. Ladder safety therefore prevails and the ladder must be:

- At the correct angle
- Secured top and bottom
- Protrude approximately 1.0 metre beyond the top of the excavation depth
- Maintained in good condition.

Excavations of any substantial length should have more than one access and egress so that alternatives are available in emergencies. Six metres or so is probably a good practice guide. It may be necessary to provide additional lifting equipment e.g. hoist in order for materials and equipment to be safely lowered into the excavation.

21.10 Are there any other hazards which may need to be considered regarding excavations?

Excavations could also be classified as confined spaces e.g. shafts and it may therefore be possible for any of the following to apply:

- Build up of toxic fumes
- Build up of toxic gases
- Lack of oxygen
- Accumulation of hazardous substances, chemicals, etc. that are heavier than air.

Even if not a confined space, gases and toxic fumes can be problems in excavations.

Sewers and drainage systems may give off hydrogen sulphide gas which can be fatal. In tolerable quantities, hydrogen sulphide gas smell like bad eggs and when this is smelt, the excavation should be evacuated, ventilated and the source of the gas identified. But in its most concentrates levels where it is breathed in, hydrogen sulphide gas is odourless – the silent killer for drainage workers.

If any work is being undertaken near sewers, drainage pipes, inspection chambers, etc., a safe system of work must be followed. Gas detection devices are essential as they monitor the concentration of various gases.

Oxygen deficiency can also be a safety hazard – often caused by oxygen being pulled out of the excavation by pressure differences, or because of chemical reactions for fumes, etc.

Carbon monoxide could build up where there is insufficient oxygen and again, carbon monoxide is a silent, odourless killer gas.

Fumes may penetrate excavations from nearby plant which is generating exhaust fumes. Diesel and petrol operated machinery should not be used near excavations unless the exhaust fumes can be directed well away from the area, or the excavation can be supplied with additional forced air ventilation.

21.11 What are the health and safety requirements regarding the demolition of structures?

CDM 2007 contains a specific Regulation on the health and safety issues relating to demolition or dismantling of a structure.

Regulation 29 states that the demolition or dismantling of a structure, or part of a structure, shall be planned and carried out in such a manner as to prevent danger or, where it is not practicable to prevent it, to reduce danger to as low a level as is reasonably practicable.

Regulation 29 goes on to state that the arrangements for carrying out such demolition or dismantling shall be recorded in writing before the demolition or dismantling work begins.

Demolition works can create major site hazards and can be the cause of death and severe injury amongst site workers and others e.g. members of the public passing by or those working in adjacent premises.

The person planning the demolition or dismantling project must consider all aspects of the task so as to prevent or reduce danger to all persons who may be affected by the works.

21.12 What are the practical requirements for managing demolition and/or dismantling works?

First, ensure that whoever is designing and implementing the demolition or dismantling works is competent to do so.

All but the most simple of demolition works will benefit from the assessment of a structural engineer and/or surveyor.

Consider what needs to be demolished, when, why and how.

Consider the environmental factors – weather conditions, wind speeds, etc.

What information has been collected regarding the structure to be demolished – i.e. information from the Health and Safety File for the building, the way the original building was constructed and the recommended sequence for demolition.

Consider the use of the building and if any persons will still be in the vicinity when demolition takes place.

Will explosives be needed?

21.13 What procedures can be put in place to manage excavations safely?

Whenever excavations are required adjacent to or in the vicinity of underground services, or when information about the location of services is incomplete, it is advisable to follow a Safe Digging Procedure as follows:

Before digging:

- Make sure that all workers involved in the digging know about safe digging practice and emergency procedures and that they are properly supervised.
- Make sure that the locator is used to trace as accurately as possible the actual line of any pipe or cable or to confirm that there are no pipes, cables or services in the way and mark the ground accordingly.
- Make sure that there is an emergency plan to deal with damage to cables or pipes. Ensure that there is a system to notify the service owners.

Excavation:

- Keep a careful watch for evidence of pipes or cables during digging and repeat checks with the locator.
- If unidentified services are found, stop work until further checks can be made.
- Hand dig trial holes to confirm the position of pipes or cables.
- Remember that plastic pipes cannot be identified with most normal locating equipment.
- Hand dig near any buried pipes or cables or use air powered excavation devices. Use spades and shovels rather than picks and forks as these are more likely to pierce cables.
- Do not use hand held power tools within 0.5 m of the indicated position of an electricity cable.
- Do not use an excavator or excavate within 0.5 m of a gas pipe.
- Treat all pipes and cables as "live" until it is known otherwise. Rusty pipes may still be in use.
- Once services are exposed, support them to prevent collapse and damage.
- Report any suspected damage to services.
- Backfill around pipes and cables with fine material and ensure it is well compacted.
- Update any existing plans with the type and location of new services or create new plans which clearly indicate all relevant information so that future users can work safely.

Work should, where possible, be away from the lines.

The lines may be temporarily switched off.

Plan what works need to be carried out in the area of overhead power cables – measure the height of the cables and specify that any long pieces of material or equipment shall be less than the height of the cables, including a safe tolerance area.

Prevent vehicle access to overhead power lines by creating barriers and restrictive areas.

22.3 Does the CDM Co-ordinator have to obtain information about the location of services from the utility providers?

The CDM Co-Ordinator has to ensure that all appropriate information is made available to those who may need it during all stages of the project.

Information regarding the location of services – especially those underground – will be invaluable to the planning of safe systems of works.

Many accidents and incidents area caused because ground workers come in to contact with underground services – a good location plan and prior identification will reduce this likelihood. The CDM Co-ordinator must advise the Client and the Designers of the need for information and could approach the utility companies for information.

Alteratnively, the CDM Co-ordinator could ensure that the Designers or Building Services Consultants have obtained the information – perhaps by using Geo-Technical and Mapping Surveyors.

22.4 Should the CDM Co-ordinator take any action if the underground or overhead services create a safety hazard?

The CDM Co-ordinator should discuss any concerns regarding the location of services with the Design Team and should ask them to co-operate and co-ordinate with each other to reduce the risks on the site.

This could include actions to review the design proposals eg. could drainage runs be relocated away from underground services or consideration could be given to relocating the services involved. This would involve the utility or supplier company.

If underground or overhead services are identified early on in the project then Designers can make sensible design discussions about where they position structures, types of foundations required, height of structures etc.

The CDM Co-ordinator could use their position to influence either the Designers or the utility/underground service providers to make changes in order to improve safety.

22.5 Do utility companies or other service providers have to provide information about their services underground?

Yes. The New Roads and Street Works Act 1991 requires undertakers to provide (subject to such exemptions as may be prescribed), keep up to date and make available for inspection, records of every item of apparatus belonging to them in the street as soon as is reasonably practicable after placing it in the street, altering its position or locating it in the course of executing other such works.

The Street Works (Records) (England) Regulations 2002 set out the form and manner of keeping records of street works in England and a Code of Practice issued by the Department for Transport gives guidance on compliance with the Regulations.

Information requests should be made as early as possible, giving clear detail of the location of the proposed works, the type of work to be undertaken, duration of works, persons responsible etc.

Telecoms companies also have to provide information about their services underground as do water companies, drainage authorities, highways authorities.

23

Safe Systems of Work

23.1 What is a safe system of work?

There is no legal definition of what constitutes a safe system of work and it will be a matter of "fact and degree" for the Court to determine.

Precedence was however set in the Court of Appeal in the 1940's when the then Master of the Rolls said:

> *"I do not venture to suggest a definition of what is meant by system. But it includes, or may include according to circumstances, such matters as physical lay-out of the job, the sequence in which work is to be carried out, the provision ... of warnings and notices and the issue of special instructions."*
>
> *"A system may be adequate for the whole course of the job or it may have to be modified or improved to meet circumstances which arise: such modifications or improvements appear to me to equally fall under the heading of system."*
>
> *"The safety of a system must be considered in relation to the particular circumstances of each particular job."*

This means that a system of work must be tailored for each individual job.

23.2 What is the legal requirement for safe systems of work?

The Health and Safety at Work Etc Act 1974 sets out specifically in Section 2 that the employer is responsible for:

> *"the provision and maintenance of plant and systems of work that are, so far as is reasonably practicable, safe and without risks to health."*

The Confined Spaces Regulations 1997 also require employers to establish a safe system of work if work and entry into confined spaces cannot be avoided.

23.3 What are the provisions for a safe system of work?

Generally, developing a safe system of work will involve:

- Carrying out a Risk Assessment
- Identifying hazards and the steps which can be taken to eliminate them
- Designing procedures and sequences which need to be taken to reduce exposure to the hazard
- Considering whether certain things or actions need to be completed before others
- Designing Permit to Work or Permit to Enter systems
- Writing down the procedure
- Training employees and others.

23.4 Is a Method Statement the same as a safe system of work?

Generally, the two are similar and a method statement is a written sequence of work which should be followed by the operator in order to complete the task safely.

This is the same as a "safe system of work" which is a sequence of events needed in order to reduce or eliminate the risks from a hazard which in itself cannot be eliminated.

Method statements are common in the construction and maintenance industries and they are often required under construction laws e.g. demolition works must always be accompanied by method statements.

The Control of Asbestos Regulations 2006 require all work with asbestos to be supported by a work plan or plan of work i.e. method statement or safe system of work.

When employees have to undertake hazardous tasks or when they have to work in hazardous environments, it is incumbent on the employer to ensure the safety of their employees and others. They must therefore decide *how* the job is to be done in order to ensure that their employees are kept safe.

23.5 What is a Permit to Work system?

A Permit to Work, or Permit to Enter, is a formal system of checks which records that the safe system of work which has been developed for the process is implemented.

The "Permit" process usually applies to hazardous areas and is commonly used for:

- Entering and working in confined spaces
- Working of electrical plant

- Working on railways or traffic routes
- Working in chemical plants
- Working in hazardous environment
- Hot works.

It is a means of communication between site management, supervisors and those carrying out the hazardous work.

Essential sections of a Permit to Work are:

- Clear identification of who may authorize particular jobs
- Limitations in respect of anyone's authority
- Clear guidance as to who is responsible fro determining the safety procedures to be followed
- Clear guidance as to what safety precautions are necessary
- Details of emergency procedures
- Information which must be relayed to site operatives
- Instructions, training and competency requirements must be specified
- The condition for which the Permit is relevant
- The duration of the Permit
- The "hand back" procedure
- Monitoring and review procedures.

There are no set forms to use – employers should devise their own but the Health and Safety Executive have a free template associated with the Approved Code of Practice and Guidance on Working in Confined Spaces.

23.6 What is a hot works permit?

Hot work is any work using open flames or sources of heat which could ignite materials in the workplace or area.

Examples are:

- Welding
- Burning
- Brazing
- Propane soldering
- Oxyacetylene cutting
- Grinding ferrous metals.

Hot works permits will be permit to work documents issued to a contractor for a specific job which has to use hot work techniques.

Hot works permits enable a safe system of work to operate during the works and ensures that all the hazards and risks associated with the task have been identified before the job starts and that the necessary control measures have been put in place.

Individuals will be held responsible for ensuring that safety measures are followed.

Permits will be issued for defined periods only and will need to be returned and signed off by a competent person.

Principal Contractors would be wise to implement hot works permits across their site and to ensure that all contractors follow the procedures.

The requirement for hot works permits on the site should be covered in the site induction procedures.

Case Study

Three men employed by a small plumbing and drainage company were called to unblock a sewer. The men lifted the manhole cover to find that the drains were about 3 metre down. The step irons looked ok and the first operative, the 17-year-old nephew of the company owner, descended into the sewer to unblock it with his rods.

When he got to the bottom of the shaft he collapsed.

His two colleagues panicked and descended into the sewer as well as rescue him, fearing that he might drown. Both men were overcome but managed to shout before passing out.

Fortunately, the Site Foreman had arrived and realized something serious was wrong. He called the emergency services.

The 17-year-old operative was dead by the time he was rescued. The other two operatives died 2 days later in hospital.

They had died of hydrogen sulphide gas poisoning – a deadly poisonous gas which, the stronger and more lethal it is, the more odourless it becomes.

The Company owner was prosecuted for health and safety offences – for failing to have a safe system of work to protect his employees whilst they were at work.

At the very least the HSE said, there should have been a safe system of work operated by a Permit to Enter/Work system.

There should have been gas monitoring/detection equipment used, the sewer sludge agitated to release any build up of toxic gas, there should have been emergency procedures, breathing apparatus, training and so much more.

24

Working at Heights

24.1 What legislation covers working at heights?

The Work at Height Regulations 2005 cover all work at height activities and came into force in April 2005.

The Regulations repeal all earlier legislation which used to cover working at heights e.g. parts of the Construction (Health, Safety and Welfare) Regulations 1996 are repealed.

24.2 What do the Regulations require employers to do?

The Regulations not only apply to employers but also to "duty holders". Duty holders will often be employers but may also cover any person who controls the way work at height is undertaken e.g. clients commissioning construction work, managing agents and building owners.

The Regulations require duty holders to ensure:

- All work at height is properly planned and organized
- Those involved in work at height are competent
- The risks from work at height are assessed and appropriate work equipment is selected and used
- The risks from fragile materials/surfaces are properly controlled
- Equipment for work at height is properly inspected and maintained.

24.3 What is work at height?

Work at height is work in any place, including a place at, above or below ground level, where a person could be injured if they fell from that place. Access and egress to a place of work can also be work at height.

Examples of work at height are:

- Working on a flat roof
- Working from a ladder

- Working near or adjacent to fragile materials
- Working at ground level adjacent to an open excavation
- Working anywhere where there is a risk of falling
- Erecting and working from scaffolding.

24.4 When is a Risk Assessment required for working at height?

Every time someone is to undertake a task where there is a risk that they could fall and injure themselves a Risk Assessment must be completed.

The Risk Assessment approach required is no different than for all other Risk Assessments completed by the employer or duty holder.

The Risk Assessment approach is to:

(1) Identify the hazard
(2) Decide who might be harmed and how
(3) Evaluate the risks and decide whether the existing precautions are adequate or whether more needs to be done i.e. identify the control measures necessary to reduce the risks of injuries
(4) Record your findings
(5) Review the assessment.

24.5 What type of issues should be addressed in the Risk Assessment?

Some examples of what to consider are:

- The work activity
- The equipment to be used
- The duration of the work
- The location where the work activity is due to take place i.e. presence of hazards such as overhead power lines, open excavations, underground services, vehicle access, people still at work, public access and so on.
- The working environment e.g. weather conditions, lighting, type of ground, slopes and so on.
- Condition and stability of existing work surfaces
- Physical capabilities of the workers e.g. vertigo sufferers, pregnancy.

24.6 What is the most important aspect of working at height that must be considered?

The most important aspect of working at height that an employer should consider is whether the work at height needs to be done at all.

The first principle of safe working at height is not to work at height at all. Consider whether there are alternative ways to do things e.g. if light fittings are high up over say, a stairwell, and access to them is unsafe due to height and risk of falls, consider whether the light fittings could be put on to winch downable mechanisms so that maintenance and cleaning can be carried out at ground level.

Work at height should only be carried out if there is really no alternative.

24.7 People carrying out work at height have to be competent. What does this mean?

Competency is the experience, knowledge and appropriate qualifications that enable a worker to identify the risks arising from a situation and the measures needed to be taken to eliminate or reduce them.

Anyone working at height needs to be trained in the hazards and risks and in what control measures are being put in place to eliminate or reduce the risks of injury. Those undertaking work at height activity must be trained in the selected system of work and on the use of any particular equipment chosen e.g. how to use mobile elevating platforms, tower scaffolds and so on.

Managers and supervisors must check that anyone carrying out work at height are trained and competent to do so.

24.8 Is it now illegal to use ladders and step ladders?

The Work at Height Regulations 2005 do not ban the use of ladders or step ladders but require consideration to be given to their use.

Ladders and step ladders should only be considered where the use of other, more suitable work equipment e.g. mobile elevating platforms, tower scaffolds are not appropriate.

Where ladders and step ladders are used they should only be used for light work of short duration.

A comprehensive Risk Assessment needs to be carried out.

24.9 What types of work equipment is suitable for work at heights?

The following types of equipment can be considered as options to reduce the hazards and risks of working at height:

- Work restraint system
- Work positioning system
- Rope access and positioning system
- Personal fall arrest system.

24.10 What is a work restraint system?

A work restraint system is a fall prevention system which relies upon personal protective equipment i.e. harness and lanyard, being adjusted to or set to a fixed length that physically prevents the person from getting to the place where they could fall.

Such a system requires strict supervision as there may be a risk that the operative will unfix their lanyard to reach more distant work locations.

24.11 What is a work positioning system?

A work positioning system is a personal fall arrest system which includes a harness connected to a reliable anchor point to support the user in tension or suspension in such a way that a fall is prevented or restricted. Examples include a boatman's chair, people working on telephone lines.

All work positioning systems must be provided with a back up system in case the primary support fails.

24.12 What are rope access and positioning systems?

This system is similar to abseiling and involves the use of two ropes each secured to different anchors. One rope is connected to a harness and the other acts as a safety back up rope. Rope access is often used to access cliff faces or the sides of tall buildings when other systems are not feasible.

24.13 What is a personal fall arrest system?

A personal fall arrest system is a fall protection system that uses a harness connected to a reliable anchor to arrest and restrict the fall and prevent the user from hitting the ground.

It usually has an energy absorbing device to limit the impact of gravity forces on the body.

24.14 What steps need to be taken to prevent objects falling from heights?

The Regulations require the likelihood of falling objects to be considered and eliminated or reduced wherever possible.

Toeboards, solid guards, debris nets, etc. should be considered.

Good housekeeping will reduce the likelihood of objects falling as tools, materials, etc. will be put away so as not to cause trip hazards and be in walkways where they can be inadvertently kicked over the edge of platforms.

If necessary tools could be tied onto harnesses or belts, especially if they have to be put down frequently.

24.15 Does the 2.0 metre rule regarding falling from heights still apply?

No. The Work at Height Regulations 2005 apply to all work at heights, no matter how far the distance of the potential fall. Injuries occur from low distance falls as well as high distance falls.

If there is a risk of a person falling *any* distance, whether off a height or into a trench or pit, then the Work at Height Regulation 2005 requires the employer, the self-employed, employees and anyone who controls the way work at height is undertaken to eliminate, reduce or control the risk of personal injury.

24.16 Are there any rules for the height of guard rails, etc.?

Guard rails must be at least 950 millimetre from the walkway or level surface or from the edge which any person is liable to fall.

Any intermediate guard rails must be positioned so that the gap between it and other means of protection does not exceed 470 millimetre.

Toeboards should be a minimum of 150 millimetre high.

24.17 What are the safety requirements for using mobile elevating work platforms?

Mobile elevating work platforms, often known as MEWPs or cherry pickers, are a much safer way of working at height than using ladders.

The hierarchy of risk control states that if a hazard cannot be eliminated a less hazardous approach should be considered. Working at heights is a hazard and using a ladder is less risky but using an MEWP is less risky than a ladder. This option should always be considered.

Safety procedures for using MEWPs include:

- Only trained operatives to use the equipment
- Equipment to be well maintained and checked
- Equipment to be suitable for the job
- The ground to be level so that the equipment is not at risk of falling over

- Access ways to be kept clear
- Overhead obstructions, especially overhead power cables, should be assessed and work programmed to avoid contact
- Safe working loads (SWL) of the platform to be considered so that they are not overloaded with operatives and materials
- Operatives to wear fall arrest equipment – e.g. harness and lanyard
- Weather conditions to be considered especially wind speeds.

MEWP: Case Study

The problem

Two operatives fell to the ground from the basket of a "cherry picker" when the basket suddenly tipped down. They both sustained severe injuries and were off work for many weeks.

As with many employees and the self-employed in the construction industry, they received no wages and both endured financial hardship. The employer lost two experienced operatives and had to hire in people from an agency at a much greater cost.

What probably happened?

The investigation showed that it was likely that the equipment was not being properly maintained and that a fastening bolt connecting one of the rods on the basket was either missing or had worked its way loose during the operation of the cherry picker.

Neither of the two operatives were wearing safety harnesses and neither had received any proper training in how to use the equipment, the hazards and risks associated with it. They also had not been trained in how to conduct a simple visual check of the equipment.

The solutions

The employer or whoever authorized the cherry picker to be used should have ensured that the equipment was safe to use and that it had had a suitable inspection by a competent person within the last 6 months – as required by the Lifting Operations and Lifting Equipment Regulations 1998.

The employer or person in control should have requested copies of their inspection certificates for the equipment.

An inspection of the cherry picker would have identified faults with the fixings.

The operatives should have had training in using the cherry picker and then they would have been declared competent to use the equipment.

The Management of Health and Safety at Work Regulations 1999 require all persons to be competent to undertake work activities.

A comprehensive Risk Assessment should have been carried out, identifying the hazards and risks associated with using the equipment. Were potential tipping hazards identified, was the weather and wind speed considered, was there a plan to reduce any hazards, etc.?

Where hazards cannot be eliminated completely, leaving the possibility of accident and injury, there should be an emergency plan in place. What steps would be taken in an emergency, who would do what, how quickly would people be rescued?

What consideration was given to personal protective equipment i.e. safety harness and lanyard? Although both operatives were inside the basket which had guard rails, there was a risk that if the basket tipped up they could fall out. They would obviously then fall from a height and the Work at Height Regulations 2005 requires that to be prevented where possible.

What were the costs?

The company could have been fined an unlimited amount for breaching the Work at Height Regulations 2005, up to £20,000 for failing to have a safe system of work and up to £5,000 for failing to have a risk assessment, failing to provide personal protective equipment, etc.

Also, as the contract was for a large public body which has to be seen to be employing competent contractors, the company was at risk of being removed from the Approved Contractor list and this would have prevented them obtaining further work.

Had the Company instigated the correct health and safety procedures it would have had to have spent approximately £300 per operative – for a safety harness and lanyard and a 1 day training course, and also about £200 to have had the equipment inspected by a competent person.

24.18 When an HSE Inspector comes to site, what will they be looking for in respect of MEWPs?

The HSE Inspector will be looking to enforce the Work at Height Regulations 2005 and the general provisions of the Health & Safety at Work Etc Act 1974.

HSE Inspectors have an inspection checklist for a wide range of job tasks and for enforcing numerous pieces of legislation.

With regard to MEWPs they will be looking to ensure that equipment and procedures meet the following:

- Should be safe plant i.e. correct type for the job, provided with adequate guard rails, toe boards
- Should be thoroughly examined in the last 6 months, properly maintained and inspected
- Should be used on a safe site (i.e. firm, level ground which is free from slopes, holes and so on) and can sustain the loads imposed by the MEWP, the operatives and material on it, etc.
- Should be segregated from site traffic to avoid collisions
- Should have a safe operator who is trained and experienced i.e. competent to use the equipment
- Should have fall protection equipment provided
- Should be no evidence of inappropriate use, unsafe practices, climbing over mid rails, leaning over the edge to the extent that there is potential to fall
- Should be appropriate anchorage for any fall arrest equipment.

Inspectors are encouraged to take a robust approach to enforcement and their guidance advises them to serve Prohibition Notices for a number of contraventions.

Prohibition Notices under the Health & Safety at Work Etc Act 1974 would be served for the following:

- No segregation of the use of MEWP from other vehicles or pedestrians e.g. street lighting works taking place with no control over traffic and so on.
- Any situation where there is a risk of vehicle impact
- Any evidence of operatives MEWPs at reckless speeds
- Any evidence that other vehicles are being driven at speed in the vicinity of MEWP
- Evidence of poor ground conditions
- Evidence of any likelihood of the MEWP over turning due to slopes, uneven ground, etc.
- Evidence of failure to consider safety issues from manhole covers, ducts, suspended floors i.e. whether they will take the weight of the MEWP
- Any evidence of collapse of any structure
- Any evidence of a MEWP's wheels or outriggers sinking into the ground or not being properly supported
- Unsafe practices being followed
- Evidence that fall protection procedures are not in place
- Evidence of inadequate anchorage points for any fall arrest system
- Evidence that the MEWP had not been properly maintained and interpreted in the previous 6 months and some evidence that competent parts were faulty.

Breaches of Prohibition Noticed carry fines of up to £20,000 per offence. If a Prohibition Notice is being ignored and a fatal or major injury occurs, there would be an unlimited fine and a risk of imprisonment.

24.19 What are some of the simple every day controls regarding safe working at heights which I can instigate on my site?

The most effective control to implement is a rigorous compliance with the requirement to guard all leading edges, drops, voids, lift shafts, holes, roof edges, etc. with suitable guard rails where there is a risk of anyone falling *any* distance. However, if there has to be a standard, it must be that anywhere from which people can fall 2.0 metres or more *must* be guarded.

24.20 What are the requirements for guard rails?

Guard rails must be substantial and properly fixed so that they are secure. Falling against a guard rail which gives way is no protection at all.
    Guard rails must:

- Have a top rail at a height of at least 950 millimetre from a level surface
- Have an intermediate guard rail at an approximate height of 470 millimetre
- Have a toe board of approximately 150 millimetre in height
- Be continuous around an opening
- Clearly visible
- Not breached by openings or missing rails, etc. e.g. at lift/hoist openings.

Guard rails must be inspected regularly by a competent person and any defects repaired immediately. Hazard warning signs must be displayed if necessary but any defect to a guard rail renders the rail unsafe and so work should be prohibited in the area until the guard rail has been repaired.

24.21 Do guard rails at the old height of 910 millimetre have to be changed to meet the new height of 950 millimetre?

No, not necessarily.
    Where existing handrails which measure 910 millimetre high are fixed in place they can remain at that height until changed.
    Any gap between the top rail and the intermediate rail should not be greater than 470 millimetre.
    The responsibility for deciding on whether the exiting guard rail height should be changed will rest with the Designer and they must complete a Design Risk Assessment.
    If the area protected by the guard rails is to have more plant and equipment placed within it, or if more frequent access for maintenance will be required,

it will be necessary to consider the effectiveness of the existing guard rails and whether, in light of new circumstances, the rail heights should be raised.

Anything which substantially changes an existing situation should be looked at as a new design and the most up-to-date requirements must be applied.

24.22 When would it be reasonable to use fall arrest systems?

Fall arrest systems are personal protective equipment which does not prevent someone from falling but if they did, it *arrests* their descent so that they do not fall all the way to the ground.

Fall arrest systems usually consist of:

- Lanyard or rope
- Harness for the body
- Hooks and couplings to connect the restraint ropes, etc. to the body harness
- Hooks or couplings to connect the lanyard or rope to the securing point e.g. eye bolt, permanent rail and so on.

Fall arrest systems should only be used when fall prevention is impossible e.g. when erecting scaffolding.

The Risk Assessment should state whether fall arrest systems are suitable control measures to reduce the risk of injury from working at heights.

Fall arrest systems may be the only safety precaution available if work is of a short duration at heights.

24.23 What needs to be considered when using fall arrest systems?

Harnesses and lanyards are made of man-made fibres and as such are subject to degradation by sunlight, inclement weather, chemicals and so on.

It is important to carry out detailed daily checks of the equipment prior to use. A visual and tactile inspection is necessary – look and feel for faults in the lanyard or harness. Choose an area in good light.

If there is the slightest doubt about the condition of any part of the fall arrest system do not use it. Label it as "defective" and put aside for further inspection.

Faults can be identified by:

- Discolouration
- Tears
- Fraying
- Nicks
- Grittiness
- Rust on metal catches
- Excessive scratching.

Harnesses do *not* prevent a fall. The person falling may not fall to the ground but will be suspended in mid air or only fall to the next level down. There is a significant risk of injury even when using a fall arrest system because people could be harmed by impact to the body when the lanyard goes taunt or when they strike against parts of the structure during the fall.

Consider the use of an energy absorbing device fitted to energy absorbing lanyards so as to reduce the risk of injury from impact loads.

Minimize the "freefall" distance. Keep the anchor as high as possible so as to reduce fall distances.

Ensure that anchor points are suitable and have been checked by a competent person. Any anchor points which a fall arrest system is attached is "lifting equipment" under the Lifting Operations and Lifting Equipment Regulations 1998 and *must* be inspected by a competent person every 6 months.

Ensure that the lanyard is not too long so that it has chance to arrest the fall before the person hits the ground.

Anyone who needs to attach themselves to a fall arrest system needs to be able to do so before they have entered the area from which they need to be protected.

Develop a system of running lines and second lanyards so that a person can unclip themselves and clip themselves on in a continuous process without being put at risk of falling.

Ensure that everyone wearing a harness and lanyard knows how to wear it properly, how to check it, adjust it, etc. Everyone should carry out their own visual pre-use checks and must themselves be satisfied about its conditions.

Never force someone to wear a harness or lanyard when they believe it to be defective.

24.24 What needs to be considered in respect of temporary suspended access cradles and platforms?

Temporary suspended access cradles and platforms are *often* used for window cleaning and external facade maintenance and cleaning. They are usually fixed at anchor points on the top of buildings and the cradle passed over the edge to "hand down" the building facade.

Many accidents happen when using suspended access cradles, mostly due to:

- Unsafe access to and from the cradle e.g. stepping over a parapet wall
- Insufficient or poorly secured counterweights and holding systems
- Failure of the cradle platform or components such as pins, bolts and so on
- Failure of winches, climbing devices, safety gear and ropes as a result of poor maintenance
- Poor erection and dismantling techniques.

Equipment must be chosen and installed by a competent person.
A comprehensive risk assessment is required.
Consideration must be given to:

- Type of weather
- Wind speeds
- Access into the cradle
- Egress from the cradle
- SWLs of cradles to include number of operatives and type and number of equipment to be used and therefore kept in the cradle
- Protection of ropes or suspension cables over parapet walls, etc. so as to prevent friction and possible fraying
- Risk of overturning
- Likelihood of suspension cable collapse and an alternative backup cable, etc.
- Guarding of the cradle
- Operating instructions
- Training and experience of the operatives.

An emergency plan must also have been devised so that if there is a failure of the system or an injury to operatives in the cradle, emergency measures can be taken for rescue, etc.

Access into the cradle should preferably be from ground level.

All access cradles should be raised when not in use so that unauthorized access is prevented, especially by children. Power should be switched off and isolated.

Access cradles must display suitable safety signs and have SWLs displayed. SWLs should be equated to understandable information e.g. number of people as opposed to kilogram weights.

24.25 What are the common safety rules which should be followed for working at heights?

Committing the following to memory and always following them will help reduce injuries for falls significantly:

- Do not work at height unless it is absolutely essential
- Make sure any working platform is secure and stable
- Check that any working platform will support the weight of workers and the necessary equipment
- Make sure any access equipment or working platform is stable and will not overturn
- Do not erect equipment on uneven ground
- Ensure that everyone has adequate working space. Overcrowded and restricted work areas contribute to accidents
- Foot any ladders or access towers or secure them to a stable structure

24.37 What is a supporting structure?

A supporting structure is:

"any structure used for the purpose of supporting a working platform and includes any plant used for that purpose"

As an example, mobile access equipment is a supporting structure, a fork lift truck will be a supporting structure, the "tower" of a tower scaffold is a supporting structure.

If an upper floor is a working platform then the walls and structure supporting the floor will be a supporting structure.

Case Study

Roof maintenance e.g. gutter cleaning: planning the task

Questions to ask include:

Height
How high is the job from the ground?

Surface
How high is the job from the ground?

Ground
What is the ground condition under the area where access equipment might need to be set up – for example, is it sloping, muddy or uneven? The access equipment you use must be suitable for the ground conditions – stable, level and not liable to fall or collapse.

If you fall, what will you fall on to? What will be the severity of injury?

Weather
Is it raining hard, or very windy? Is it icy, snowing or generally wet?

Task
What tools will you use to clean the gutters? How will you manage the debris? How will you get the tools to where you need them?

When looking at what you need to do the job, think about the following table.

From the ground	Can you do the work from the ground?	Yes/No
From the roof	Is the roof above the gutter accessible?	Yes/No
	Is the roof strong enough to work from?	Yes/No
	Does the roof have guard rails or other equipment that will prevent a fall? If no, can this be installed?	Yes/No
From a platform	Can you do the work from a mobile elevated platform (MEWP) or tower scaffold?	Yes/No
Fall Protection	Do you need personal fall protection to allow safe access?	Yes/No

(Continued)

From a ladder	Is the work low risk and short duration?	Yes/No
	Do you have a ladder that will reach the area?	Yes/No
	Can you secure the ladder safely?	Yes/No
	Can you or your workers use the ladder safely?	Yes/No

Consider what can go wrong?

Could the access equipment:

- Slip
- Fall over/topple over
- Block means of escape from the building
- Block means of access
- Obstruct footpaths, etc.
- Be knocked into by passers-by
- Be too big awkward to be erected safely?

Could the task to be undertaken:

- Cause debris to fall on to passers-by
- Cause tools to fall on to passers-by
- Obstruct others carrying out different work near by
- Interfere with any safety equipment.

Could the operative undertaking the task:

- Be incompetent/inexperienced
- Lack training
- Misunderstand the hazards
- Take short cuts, rush the job
- Not know how to use the tools, etc.

Tips for safe working

- Deal safely with the gutter debris. It is best to bag it in small quantities and lower it slowly to the ground – if there is a risk of falling debris, make sure no-one comes into the area below the work
- Take frequent breaks, especially when working from a ladder – do not work from a ladder for longer than 30 minutes at a time
- If you are using a ladder keep three points of contact wherever possible
- Make sure the people who will be doing the job have the right skills, experience and training to use the equipment safely and have been consulted about the right equipment to use
- If you are hiring access equipment, make sure you know how to install and dismantle it safely – ask the hirer for instructions or assistance if you need them

Case Studies

A maintenance worker, aged 38, plunged 30 feet (10 m) to his death after falling through an asbestos roof, an inquest heard. He went onto the roof of a building to repair panels which had been damaged in a break-in, after being told by his boss to "sort it".

A scaffolding company was fined heavily after a court heard how a scaffolder was killed when he fell from a third storey window ledge to the street below.

The scaffolder was removing a large hoarding from the front of a building. The scaffolder climbed through the window on the third floor in order to gain access to a narrow ledge so that he could stand on it to reach the hoarding brackets.

He lost his balance and fell 12 feet (4 m). to his death.

The Company was prosecuted for:

- Failing to have a safe system of work
- Failure to carry out risk assessments
- Failure to provide employees with adequate information, instruction and training.

The HSE Inspector said that an MEWP should have been used. The procedure adopted was inherently unsafe.

A Store Assistant who was working in a retail store died from head injuries after falling from a ladder, an inquest heard.

He was using the ladder to remove a bike part from an upper shelf in the store room when he fell from it.

A colleague found him on the concrete floor on all fours with blood coming from his head.

"He was across the ladder with a cycle against him. He was bleeding from the head, he was dizzy, delirious and started being sick. I rang for an ambulance immediately," he said.

The storeman underwent brain surgery in a local hospital but died 4 days later.

The inquest jury returned a verdict of accidental death.

The Health and Safety Inspector said he had investigated the scene and the ladder being used lacked proper feet and had been used at an incorrect angle.

His injuries were severe and were described in court as:

"He had an abrasion and swelling on his head and there was fresh blood leaking from his right ear. He had a brain scan and suffered a fit while in the machine. The results of the scan showed serious head injuries, two skull fractures, bleeding around brain tissue and significant bruising in two parts of the brain."

25

Electrical Safety

25.1 Is it against the law to use mains electricity on a construction site?

No. The law does *not* state that you cannot use 230 volt electrical supply on a construction site. You could do so if you manage the risks adequately.

However, 230 volt electricity is a major hazard and has great potential to kill and cause injury to people if used unsafely or in conditions which are unsafe.

Managing health and safety is about eliminating or reducing hazards to acceptable levels.

Requiring contractors to use 110 volt power tools or 12 volt power tools reduces significantly the likelihood of injury from the electrical supply.

It is therefore a safe system of work to use 110 volt or 12 volt power tools.

25.2 What safety steps must I take if 230 volt is the only available power on site?

If power tools have to operate on 230 volt then the following are required as an absolute minimum:

- Detailed Risk Assessment
- All equipment, cables, leads in good repair
- Use of residual current devices (or trip devices) which operate at 30 milli-ampere with no time delays
- Regular daily checks
- Avoidance of hazardous environments e.g. eliminate damp conditions, water, dust, use of chemicals and so on.

All equipment must be connected to Residual current devices (RCDs), or if agreed by a competent electrician, an RCD can be fitted to the main power supply so that it protects the whole of the electrical system.

Risk Assessments must be constantly reviewed and up dated should conditions on-site change.

Doubly insulated equipment should be used where possible.

RCDs must be installed with great caution as a poorly installed one will create more hazards than it solves.

RCDs must be:

- Kept free of moisture
- Kept from dirt
- Protected from vibration
- Protected from mechanical damage
- Properly installed and enclosed
- Properly protected with sealed cable entries
- Checked daily by using the test button
- Considered as a safety device and not the complete eliminator of a hazard.

Tools must only be connected to sockets protected by RCDs.
Cables and leads supplying equipment running on 230 volt must:

- Be protected from damage
- Kept at high levels
- Protected inside impact resistant conduit.

For additional safety use armoured protected flexible cables/leads.

25.3 What should the Site Agent do if a contractor comes on to site and wants to use 230 volt equipment when the site operates on 110 volt?

The first response will be to refuse him access to the site on the grounds that he will not be following the site rules.

However, a more realistic approach needs to be taken as turning the contractor away may affect the programme.

A full review with the contractor is necessary i.e.:

- What needs to be done
- Who needs to do it
- Where does it need to be done
- Where are the Risk Assessments
- Where are the Method Statements
- What work may affect others in the area
- What controls for safety does the contractor propose using e.g. RCDs
- How does he propose to manage the hazard of 230 volt tools and cables
- What Permit to Work methods does he propose
- Can the work area be isolated from others so as not to expose others to unnecessary hazards.

If the Risk Assessments and Method Statements demonstrate that a safe system of work can be operated, that the risks to *all* operatives can be controlled, that the work is of relatively short duration then is *may* be possible to allow the work to go ahead.

Using 230 volt on a construction site is a serious hazard which poses great risk of electrocution, electric shock, electric burns and fire, but it is *not* against the law and provided the risks can be managed it may be permissible.

However, best practice and all safety guides advocate the use of 110 volt power tools or cordless 12 volt power tools.

25.4 Does all electrical equipment have to have an inspection and test certificate?

The Electricity at Work Regulations 1989 requires the "duty holder" (usually the employer) to ensure that electrical equipment is:

"maintained in a safe condition"

The Regulations do not specifically require elaborate testing and inspection but if these steps are needed to ensure that the equipment is safe then they must be carried out.

A simple, and usually, effective, procedure to check electrical equipment for faults and damage is ... visual inspection. Approximately 95% of faults or damage can be detected by carrying out a thorough visual inspection.

All operatives should be trained on what to look for when carrying out a visual inspection and everyone should be encouraged to use the techniques *every* time they use electrical equipment.

Before an electrical hand tool, RCD, lead or cable extension, etc. is used, check that:

- No bare wires are visible
- The cable covering is not damaged
- The cable cover is free from cuts and abrasions
- The plug is in good condition
- The plug pins are not bent, missing, the casing cracked, etc.
- There are no joined cables with insulating or other tap – any cable connectors must be proprietary
- The outer cable covering (sheath) is gripped properly where it enters the plug or equipment e.g.. no coloured wires showing
- The equipment itself is in good condition with no parts missing, screws missing, casing cracked, etc.
- There are no overheating marks, burn or scorch marks around the plug, socket, equipment
- The RCD is working correctly.

Some equipment will require more definitive evidence that it has been formally inspected and a certificate is a good record.

Equipment which has a high risk of becoming unsafe through misuse should be inspected and tested, usually suing a PAT test. This test will check the earth continuity.

If equipment is going to be used in hazardous areas e.g. confined spaces, it should be inspected and tested regularly, with suitable records kept.

If equipment is tapped to earth it should be regularly checked to ensure that the earth wire is intact and connected. Abuse of 110 volt tools by multi-users will often cause wiring to come loose. This may not be seen by visual inspection. A formal process of combined testing would be sensible.

Top Tips

Equipment	Voltage	User checks	Formal visual inspection	Combined inspection and testing
Battery operated power tools	Less than 25 volt, usually 12 volt	No	No	No
Torches	Less than 12 volt	No	No	No
25 volt portable hand lamps	25 volt, powered from transformer	No	No	No
50 volt portable hand lamps	50 volt, secondary winding, centre tapped to earth	No	No	Yearly
110 volt portable hand tools, extension leads, site lighting	Max 55 volt tapped to earth	Weekly	Monthly	Before first use on site, then 3 monthly
230 volt portable and hand held tools	230 volt mains supply	Daily/ every shift	Weekly	Before use and then monthly
230 volt portable floodlighting, extension leads	230 volt	Daily	Weekly	Before first use, monthly
230 volt "fixed" equipment e.g. lifts, hoists, permanent lighting	230 volt	Weekly	Monthly	Before first use and then 3 monthly
RCD's portable		Daily, every shift	Weekly	Before first use and then monthly
RCD's fixed		Daily	Weekly	Before first use and then 3 monthly
Mains equipment in site offices e.g. fax machines, photocopiers	230 volt	Monthly	6 monthly	Before first use, then yearly

Source: HSE : HSG 150.

25.5 What precautions which need to be taken when using electrical equipment in hazardous areas?

Hazardous areas will include:

- Confined spaces
- Flammable atmospheres e.g. paint shops
- Chemical plants
- Areas subject to dust and particle accumulation.

Wherever possible electrical equipment should *not* be used in hazardous areas.

Before any work is carried out in these areas a PERMIT TO WORK should be issued. This will detail all of the safety precautions to be taken for working in the area, including which type of electrical power tool is appropriate.

In any hazardous environment, 12 volt portable tools are safer to use than any other.

25.6 Are there any special precautions to take in respect of health and safety when commissioning and testing of fixed plant and equipment takes place prior to project completion?

Equipment will need to be commissioned during the course of the development works and this is likely to involve the use of 230 volt mains power supply.

There is a health and safety risk when using a 230 volt supply on a construction site and the following procedures should be adopted:

- The Principal Contractor should be required to produce an agreed plant commissioning programme which will incorporate information about the site and from other sub-contractors/nominated contractors who require plant and equipment to be commissioned.
- All sub-contractors will be informed about the commissioning of plant and equipment and the Principal Contractor shall display a series of notices on either the individual equipment or on the entrance to the area which advises everyone that the equipment is now live and connected to 230 volt.
- Secure means of isolation shall exist for each part of the installation on which work is carried out. Padlocks and keys should be clearly identified and held by the Principal Contractor.
- Circuits which are not in use, i.e. used for commissioning and thereafter not required until the site is handed over, should be locked off and doors to switch rooms, fuse boxes and so on should be locked shut with keys kept by the Site Agent.

- All electricians should be suitably trained to work with high voltages and obviously the Electricity at Works Regulations 1989 must be complied with.
- Each employer will be responsible for his own employees and must ensure suitable information, instruction and training has been provided, that their employees are competent to do the job, etc.
- The Principal Contractor will be responsible for ensuring that site-safety rules are followed and that all contractors and sub-contractors are kept informed of all site-safety matters.

Mechanical plant and equipment e.g. air handling units, lifts and escalators will be operating from mains voltage whilst the site is still a construction site. The hazards and risks of 230/240 volt on site must therefore be properly and safely managed.

All cables, ductwork, etc. carry 230/240 volt power must be labelled with a hazard warning sign (yellow and black) as follows:

"WARNING : LIVE 230/240 V POWER"

"DO NOT TOUCH"

or similar

All electrical cables running adjacent to work areas *shall* be encased in armour plated conduit (or other recognized material) and sufficiently and visibly labelled:

"LIVE ELECTRICS"

The Principal Contractor shall ensure that *all* operatives on site are aware that the mechanical and electrical equipment may be operational. Tool box talks and notices displayed in mess rooms will provide timely reminders.

- Adhesives
- Paints
- Wood dusts
- Welding fumes
- Carbon deposits/soot
- Acid cleaners
- Detergents/degreasers
- Pesticides.

Details from the suppliers or manufacturers must be obtained for all substances planned to be used on site, or brought in by others to be used on site.

Remember, whilst the operatives in the immediate vicinity of the hazardous substance may be protected by PPEs, others working some distance away could be affected by the substances, fumes, vapours, mists or dust.

Case Study

Two lift engineers were working in the lift shaft, welding the metal frame. They were adequately protected by PPE and had a local exhaust ventilation system which removed the welding fumes. However, unbeknown to them, the exhaust extract was faulty because it had not been subjected to regular inspection and the welding fumes were leaking into the roof void above the lift shaft in which two operatives were laying electrical cables. Both operatives in the roof void were overcome by fumes and were found when the Site Agent checked on progress of works.

26.5 How do hazardous substances affect health?

All types of hazardous substances will affect the body in different ways, depending on how the substance enters the body, known as the route of entry.

26.5.1 Inhalation

Inhalation of a hazardous substance accounts probably for the majority of deaths associated with using substances.

Inhalation of dusts, fumes, mists, vapours will affect the health of an individual in different ways and at different speeds.

Acute exposure refers to the immediate onset of symptoms i.e. the exposure is so great that an immediate adverse reaction happens e.g. unconsciousness, acute respiratory failure, heart attack and so on.

Chronic exposure is exposure to substances over a prolonged period of time where the harmful substance accumulates in the body over that period of time – often little and often exposure.

Lung diseases occur often because of persistent and long-term exposure to the hazard e.g. miners and coal dust.

26.5.2 Ingestion

Substances once swallowed will enter the stomach and intestines and pass into the blood stream. When this happens the toxic substances are transported around the body to all major organs.

Ingestion mostly occurs from "hand to mouth" contact i.e. eating or drinking with contaminated hands. Also, if drops of the harmful substance or dust, etc. are accidentally transferred on to the mouth and the person wipes their mouth they could transfer the substance onto the tongue where it is swallowed, etc.

26.5.3 Absorption through the skin

Certain substances and micro-organisms can pass through the skin which acts as a protective barrier to underlying tissue and the blood stream.

Substances can therefore enter into tissue and the blood stream and then travel around the body to vital organs. Some substances are so hazardous that only minute droplets are needed to come into contact with the skin e.g. Ricin (a common terrorist chemical weapon) in order to cause multiple organ failure and death.

Any cuts or grazes to the skin, abrasions, etc. will increase the risk of absorption of a chemical through the skin.

Sometimes the skin does not allow complete transference of the substance into underlying tissue and the blood stream and an allergic reaction takes place. This manifests itself in visible skin irritation, rashes, blisters, burns, etc.

The most common parts and organs of the body affected by hazardous substances are:

- Lungs
- Liver
- Bladder
- Skin
- Brain
- Central nervous system
- Circulatory system
- Reproductive system
- Urinogenital system.

The most common illnesses suffered by construction workers are:

- Lung diseases e.g. pneumoconiosis
- Lung cancer
- Silicosis
- Dermatitis

- Asthma
- Leptospirosis
- Legionnaires disease.

26.6 What are occupational exposure limits or standards?

For a number of commonly used hazardous substances, the Health and Safety Commission has assigned occupational exposure limits (or standards) (OEL/OES) to help define what is adequate control.

OEL are set at levels which will not damage the health of employees exposed to the substance by inhalation, day after day.

Where a substance has an OEL the exposure of employees to the substance must legally be reduced to the OEL level.

26.7 What are maximum exposure limits?

Maximum exposure limits (MEL) are set for substances which can cause the maximum amount of health damage. These substances usually cause life-threatening illnesses such as cancer, asthma, severe industrial dermatitis, respiratory conditions, etc.

Substances which have an MEL must be used only if there is no alternative and exposure must not exceed the stated limit over the given exposure time – usually no more than 10 minutes.

Employers should avoid the use of all substances with an MEL – find an alternative. Designers and specifiers should be aware of the risks from products which have an MEL and avoid them, thus providing lower risk for the contractors.

26.8 What is health surveillance?

Health surveillance is required under certain circumstances and requires the employer to assess the health of his or her employees regularly. If employees are exposed, for instance, to a substance which causes skin irritation, then it may be necessary to check the condition of hands and arms by visual examination from time to time.

Health surveillance allows an employer the opportunity to monitor the effectiveness of the control measures in place.

If employees are exposed to breathing in fumes or dust, then routine lung tests or blood tests can be used.

Health surveillance can be carried out by a medical doctor or occupational nurse, or an employer can carry out simple assessments and refer to experts for advice.

26.9 Does the Principal Contractor have to arrange for health surveillance for operatives on site?

If the operatives on site are employees of the Principal Contractor then the Principal Contractor may well have the duty to organize health surveillance.

But if the operatives are employees of other contractors or are self-employed then they themselves will be responsible for complying with the COSHH Regulations and arranging health surveillance as necessary.

Principal Contractors may be responsible for arranging for health surveillance for all operatives on site if the potential health hazards are site wide.

Principal Contractors should assess the procedures that contractors have for health surveillance as part of their competency and assessment investigations.

Occupational ill-health is a common problem on construction site and a Principal Contractor could implement a site-wide scheme to carry out health checks on all operatives for skin dermatitis. This could be a simple process of checking hands and forearms of all workers leaving site one day.

Information should indicate to the Principal Contractor whether appropriate PPE is being worn and whether it is working well.

If operatives are displaying signs of skin irritation then the Principal Contractor is in a good position to investigate the use of materials, the application practices, the use of PPE, etc.

Pro-active health surveillance by whoever undertakes will always highlight trends in potential health risks and good, preventative action can then be taken.

26.10 What are the requirements for a COSHH Assessment?

Regulation 6 of the COSHH requires employers to carry out a "suitable and sufficient" assessment of the risks to health from using the hazardous substance.

"Suitable and sufficient" does not mean absolutely perfect but the guidance which supports Regulation 6 lists a number of considerations which must be taken into account during the process.

The first step in providing a suitable and sufficient COSHH Assessment is to ensure that the person carrying it out is *competent*.

Regulation 12(4) of COSHH requires any person who carries out duties on behalf of the employer to have suitable information, instruction and training.

A competent person does not necessarily need to have qualifications as such, but they should:

- Have adequate knowledge, training, information and expertise in undertaking the terms "hazard" and "risk".
- Know how the work activity uses or produces substances hazardous to health.

- Have the ability and authority to collate all the necessary relevant information.
- Have the knowledge, skills and experience to make the right decisions about risks and the precautions that are needed.

The person carrying out the assessment does not need to have detailed knowledge of the COSHH Regulations, but needs to know who and where to go to obtain more information. They need to be able to recognize when they need more information and expertise. As the saying goes "a little knowledge is a dangerous thing".

The COSHH Assessment process can be broken down into *seven steps* as follows.

26.10.1 Step 1: Assess the risks

Identify the hazardous substances present on the site, or intended to be used on the site.

Consider the risks these substances present to your own employees and all others.

List down all of the substances likely to be used. If the substances are on site, read the labels and look for the hazardous warning symbols as follows:

Symbol	Meaning	Description of risks
	Toxic (T) Very toxic (T+)	Toxic and harmful substances and preparations posing a danger to health, even in small amounts. If very small amounts have an effect on health the product is identified by the toxic symbol.
	Harmful (Xn)	These products enter the organism through inhalation, ingestion or the skin.
	Highly flammable (F) Extremely flammable (F+)	(F) Highly flammable products ignite in the presence of a flame, a source of heat (e.g. a hot surface) or a spark. (F+) Extremely flammable products can readily be ignited by an energy source (flame, spark etc.) even at temperatures below 0°C.
	Oxidising (O)	Combustion requires a combustible material, oxygen and a source of ignition; it is greatly accelerated in the presence of an oxidising product (a substance rich in oxygen).
	Corrosive (C)	Corrosive substances seriously damage living tissue and also attack the other materials. The reaction may be due to the presence of water or humidity.
	Irritant (Xi)	Repeated contact with irritant products causes inflammation of the skin and mucous membranes etc.
	Explosive (E)	An explosion is an extremely rapid combustion. It depends on the characteristics of the product, the temperature (source of heat), contact with other products (reaction), shocks or friction.
	Dangerous for the environment (<<N)	Substances: highly toxic for aquatic organisms, toxic for fauna, dangerous for the ozone layer.

Obtain safety data sheets from the supplier or manufacturer.

A safety data sheet must be provided by the manufacturer or supplier and it must give information on, amongst other things, common usage, constituent ingredients, exposure limits, PPE, emergency procedures, first aid, spillage precautions.

If Sub-Contractors are supplying the hazardous substances then they must be required to submit manufacturers data sheets together with their COSHH Assessments.

The COSHH Regulations 2002 introduced a revision to the information needed for a COSHH Assessment and now *all* COSHH Assessments must have attached to them the product's safety data sheet.

Consider the risks of the hazardous substances identified to people's health. Remember, it is all people and not just employees. Again, the 2002 COSHH Regulations clearly state that the effect on health of a hazardous substance to "other persons" must be clearly considered.

The assessing of risk to health from using the hazardous substance is one of judgement. Sometimes it is not possible to know for sure what level of exposure will be harmful. In these cases, it is preferable to always err on the side of caution.

Some questions to consider are:

- How much of the substance is in use
- How could people be exposed
- Who could be exposed to the substance and how often
- What type of exposure will they have
- Could other people be exposed to the harmful substance.

26.10.2 Step 2: Decide what precautions are needed

The first responsibility for an employer is to eliminate the risk of using a harmful substance. This is an effective precaution to take but may not always be possible.

Next, consider whether an alternative substance could be used which is less hazardous than the original substance. If there is something on the market which does the job more safely then it should be used.

If the substances cannot be eliminated or substituted, then it must be used with *suitable controls* so as to reduce any level of exposure to an acceptable limit.

Suitable controls may be:

- Changing the way the work is done e.g. painting instead of spraying
- Reducing concentrations of the substance
- Modifying the work process e.g. to reduce exposure time
- Reducing the number of employees and others exposed to the substance
- Adoption of maintenance controls and procedures
- Reduction in quantities of substances kept on site

26.10.6 Step 6: Carry out health surveillance

Health surveillance is defined as:

"an assessment of the state of health of an employee, as related to exposure to substances hazardous to health, and includes biological monitoring".

Health surveillance will be necessary in the following circumstances:

- Any exposure to lead fumes
- Exposure to substances which cause industrial dermatitis
- Exposure to substances which may cause asthma
- Exposure to substances of recognized systemic toxicity i.e. substances that can be breathed in, absorbed through the skin or swallowed and which affect parts of the body other than where they enter.

Any health surveillance carried out should be recorded and records should be kept for at least 40 years. It is for the *employer* to decide whether health surveillance is necessary.

26.10.7 Step 7: Information, instruction and training

All employees expected to use or to come into contact with substances hazardous to health shall receive suitable information, instruction and training.

The Site Agent should ensure that all contractors have adequate records for training their operatives in the health and safety risks of using hazardous substances.

The Site Agent should ensure that the site induction training covers the use of hazardous substances on the site and the control measures to be followed.

All operatives, visitors, contractors, etc. are entitled to see the COSHH Assessment and the attached safety data sheet.

Information may be way of safety notices, tool box talks, site rules, etc.

Instruction generally covers one to one exchanges on how to do something, what to use, what PPE to wear, etc. The health effects of using the hazardous substance should be clearly discussed.

It is useful to keep records of any exchange of information on COSHH in the site records book.

26.11 What are the health hazards associated with lead?

Lead is a major health hazard and is controlled by its own regulations:

"Control of Lead at Work Regulations 2002".

Managing the risks from lead on site can be quite complex, although the general principles of COSHH broadly apply.

Any exposure to lead, lead fumes or vapours, dusts, etc. must be subject to a "suitable and sufficient" risk assessment.

Specific controls are required if employees are going to be exposed to "significant exposure" i.e.:

- Exposure exceeds half the occupational exposure limit for lead
- There is substantial risk of an employee ingesting lead
- There is a risk of the employees skin coming into contact with lead alloys or alkyls.

If exposure is deemed to be "significant" then an employer must:

- Issues employees with protective clothing
- Monitor lead in air concentrations
- Place employees under medical surveillance
- Use Risk Assessments to determine control measures
- Identify any other measures to reduce exposure and comply with the Regulations.

26.12 How is the level of silica dust determined?

If a significant level of stone cutting, cement drilling/cutting, etc. is to be undertaken on site it will be highly likely that the airborne dust levels of silica will be high.

The Principal Contractor should arrange for air monitoring to be undertaken by a competent person so that levels can be determined.

Results can be analysed in the laboratory and in some instances, immediate indication can be given from electronic sampling equipment.

Levels in excess of the MEL will need to be reduced. Generally, levels greater than 0.1 milligram per cubic metre can be regarded as harmful and significant and control measures will have to be put in place to reduce levels of exposure.

Air monitoring by competent persons should also indicate what type of control measures will reduce the dust levels e.g. local exhaust ventilation.

26.13 What are the hazards associated with silica?

Crystalline silica is present in substantial quantities in sand, sandstone and granite and often forms a significant part of clay, shale and slate. It is also found in chalk, limestone and other rocks and stones.

Crystalline silica can cause lung cancer and other lung diseases.

The health hazards of silica come from breathing in the dust. Operatives and members of the public can be exposed to inhaling the dust.

Not only could operatives undertaking activities directly involving crystalline silica be exposed to the hazards but so also could operatives in the vicinity.

Dust travels around site and can be inhaled by all persons although obviously concentrations are reduced the further away the dust travels from its source.

Activities which can expose workers or others to exposure to dust are:

- Stone masonry
- Façade renovations
- Demolition
- Sandblasting or stone cleaning
- Concrete scabbling
- Concrete cutting
- Concrete drilling
- Tunnelling.

The use of power tools will increase the volume of dust which can be created during the activity.

Exposure is intensified in confined spaces – the dust cannot dissipate and greater quantities are inhaled.

Breathing in fine particles of silica dust can cause silicosis. Dust particles will be microscopic.

26.14 What are the symptoms of silicosis?

Silicosis is the scarring of the lung tissue which leads to breathing difficulties.

Exposure to very high concentrations over a relatively short period of time can lead to acute silicosis resulting in rapidly progressive breathlessness and death within a few months of onset.

Usually, constant exposure to relatively low levels of silica dust results in progressive silicosis. Time lengths vary but can be several years. Prolonged exposure causes fibrosis – a hardening or scarring of the lung tissue – which results in the loss of lung function. Victims are likely to suffer shortage of breath and will find it impossible to walk even short distances once the disease has taken hold.

The condition is irreversible and the condition continues to develop even when exposure to the dust has ceased.

Heart failure is a common cause of death.

Sufferers often are unable to enjoy retirement and are probably not aware of the damage they are doing to their health during their working years.

26.15 Are there any safe limits for exposure to silica dust?

Silica has been assigned an MEL of 0.3milligram per cubic metre of air.

Exposure to respirable silica dust must be reduced to as low as is reasonably practicable and always to limits below the MEL.

Exposure even at limits below the MEL should be short and infrequent.

26.16 Is a COSHH Assessment necessary for silica?

Yes.

It will generally be for the Principal Contractor to complete the COSHH Assessment for exposure to silica dust across the whole site.

If the works to be undertaken are carried out by a specialist contractor and are contained in a localized area, then the specialist contractor could complete the COSHH Assessment.

The COSHH Assessment must be reviewed by the Principal Contractor in order to assess whether site wide preventative or control measures need to be implemented.

The COSHH Assessment should evaluate the risk of exposure, describe the precautions necessary to control the exposure and set out how the control measures are to be monitored, supervised and maintained.

Examples of typical levels of silica exposure in some common construction activities.

Activity	Control measures	Exposure	Improvements required*
Drilling in poorly ventilated undercroft	• No dust suppression • No extraction • No forced ventilation • Inadequate respiratory protective equipment (RPE)	**HIGH–300 times the MEL**	• Fit water suppression or dust extraction to drilling equipment • Provide appropriate PPE • Ensure correct use of RPE
Drilling onto brickwork under arch blocked at one end	• Primitive extraction by fan and airbag • Disposable face masks worn	**HIGH-5 times the MEL**	• Fit water suppression or dust extraction to drilling equipment • Provide appropriate RPE • Ensure correct use of RPE
Using jackhammers to break out concrete in large open indoor area	• Limited ventilation • No dust suppression • No local exhaust ventilation • No RPE in use	**MEDIUM-within the MEL but double toe level regarded as reasonably practicable**	• Wet down concrete and rubble

(Continued)

Activity	Control measures	Exposure	Improvements required*
Chasing out cracks in screeded cement floor in large open indoor area	• PRE provided but not worn properly • Breathing zone of worker crouching over grinder very close to source of dust	**HIGH-6 times the MEL**	• Attach dust extraction to grinder • Wet down ahead of chasing • Provide appropriate RPE • Ensure correct use of RPE
Chasing out mortar between bricks prior to re-pointing	• Ineffective extraction fitted to hand-held electric grinder • RPE correctly worn but not to correct standard	**HIGH-21 times the MEL**	• Attach dust extraction to grinder • Provide appropriate RPE • Ensure correct use of RPE
Cutting paving kerb (33% silica) in open area	• Petrol driven saw not fitted with water spray or local exhaust ventilation	**HIGH-12 times the MEL**	• Provide effective water suppression system to saw
Cutting blue brick (32% silica) in open area	• Petrol driven saw not fitted with water spray or local exhaust ventilation	**HIGH-5 times the MEL**	• Provide effective water suppression system to saw
Cutting breeze block (3% silica) in open area	• Petrol driven saw not fitted with water spray or local exhaust ventilation	**HIGH-twice the MEL**	• Provide effective water suppression to saw
Cutting window openings in concrete wall with wall saw/cutting concrete with floor saw	• Water suppression on saw used	**LOW-well below the MEL and also below the level regarded as significant**	
General clearing and removing rubble	• Hand sweeping with brush	**HIGH-twice the MEL**	• Damp down rubble before clearing • Use mechanical means to sweep up • Provide appropriate RPE • Ensure correct use of RPE

(Continued)

Activity	Control measures	Exposure	Improvements required*
General clearing and removing rubble	• Use of mechanical sweeper with rotating brushes and vacuum extraction	**MEDIUM-within the MEL but double the level regarded as significant**	• Provide appropriate RPE • Ensure correct use of RPE
Concrete crushing from demolition job for use as hard core	• Machine with enclosed cab • Water jets fitted	**LOW-well below the MEL and also below the level regarded as significant**	

* To reduce exposure to below the MEL and so far as is reasonably practicable.
Source: Reproduced from HSE: Construction Information Sheet No. 36.

26.17 What are the health hazards associated with cement?

Cement can cause ill-health mainly by:

- Skin contact
- Inhalation of dust
- Manual handling of sacks/bags.

Wet cement causes skin irritation and/or burns.

Skin irritation is usually termed dermatitis and the symptoms are red scaly skin, itching, soreness and cracking of the skin.

Some people can be allergic to the properties of cement. Allergic reactions are usually demonstrated by the same skin symptoms as irritation.

People can become sensitized to cement over a period of time – particularly older workers.

Cement dust is a major health hazard and breathing in the fine dust can irritate and clog up the lungs. In the short term, or for short exposures, the nose and throat will be irritated.

Cement dust can also contain silica – which is itself a health hazard.

Manual handling issues are associated with the weights of cement although the manufacturing industry have generally limited the weight of bags to 25 kilogram.

However, manual handling issues can be caused when sites use the on-site mixing silos and generate their own wet cement mix to be moved around site. There may be a tendency to overload wheelbarrows, carry bags, etc. so that the mix is moved quickly.

26.18 What are some of the control measures which need to be put in place to control exposure to cement?

The first step is to look at the possibilities of eliminating the need for cement or substituting the need for cement r substituting for a less hazardous substance.

Easier said than done! So, the control measures which will be practicable are ones which limit exposure to the hazardous substance.

Avoid direct skin contact with both dry and wet cement.

Issue gloves to be worn by all operatives.

Provide easily accessible wash stations with clean, warm running water, soap and towels.

Ensure that operatives wear long-sleeved garments and long trousers. Reduce exposure to the skin.

If on-site silos are used, ensure that there is no need to directly touch wither the dry material or the wet prepared cement.

Avoid the risk of anyone breathing in dust when mixing cement. Issue suitable respiratory masks to individuals, or better still, mix the cement in one controlled area and provide local exhaust ventilation.

Manual handling injuries are prevented by providing appropriate mechanical aids such as wheelbarrows, sack trucks, trollies, etc.

Ensure your supplier is providing bags which are no heavier than 25 kilogram.

Remember – the health hazards of cement accumulate over time – exposure which is little and often but over prolonged periods i.e. years is more harmful than a short sharp burst of airborne dust.

26.19 What are the health hazards from solvents?

Solvents are chemical substances used as carriers for surface castings such as paints, varnishes, adhesives, fungicides, pesticides and perhaps other specialist cleaning products.

Common solvents are:

- White spirit
- Xylene – found in paints, lacquers, adhesives
- I-butanol – found in resins, paints.

Many products will contain mixtures of solvents.

Solvents can cause ill-health through:

- Inhalation of vapours as the solvent dries.
- Skin contact – some solvents are absorbed through the skin. Prolonged or repeated contact with liquid solvents can cause industrial dermatitis.

- Eye contact – causing irritation, inflammation and potentially severe eye damage.
- Ingestion through contaminated food, drink and smoking. "Hand to mouth" contact is a common route – solvent on the hands, eat something = ingestion of solvent.

It is not uncommon for people to drink solvents – often inadvertently because the liquid has been put in an old drinks bottle and has not been re-labelled.

Poorly labelled hazardous products on a construction site are a major issue and indicate overall poor safety management practices.

The common symptoms of exposure to solvents are:

- Skin irritation
- Eye irritation
- Lung irritation – difficulty breathing
- Headache
- Nausea
- Dizziness and light headedness
- Impaired co-ordination.

Once people are exposed to the hazardous vapours of the solvent they may become more accident prone because co-ordination has deteriorated – for instance, the inhalation of a vapour cloud of solvent whilst working from a ladder could cause the person to fall off the ladder, sustaining more serious injury from the fall than from the exposure to the vapour.

Loss of concentration is common and reactions slow down. Intoxication feelings can be experienced (the reasons why some people "glue sniff" vapours which is inhaling solvent vapours).

High exposures to solvent vapours can have an immediate impact on consciousness and can cause death.

Solvent exposure is obviously increased when the work area is confined or poorly ventilated.

Large spillages can be a cause of wide spread vapour inhalation.

26.20 How is exposure to solvents controlled?

Work with solvent-based products is governed by the COSHH Regulations 2002 and exposure to the hazardous substance has to be assessed, prevented, reduced or controlled.

Following the hierarchy of risk control takes us first to step one:

- Prevent exposure – does the solvent-based substance have to be used at all? Is there a non harmful equivalent?

Heatstroke is life threatening and urgent treatment by a doctor is very important. While waiting for medical help to arrive, cool the patient as quickly as possible. Soaking the persons clothing with cold water and increasing air movement by fanning can do this. If the person is conscious, give them water to drink.

26.29 What can the Principal Contractor do to reduce the likelihood of sunburn & heat stroke?

The Principal Contractor and all other employers should make sure that all operatives are trained in ways to reduce the risk of sunburn, heat stress and heatstroke. Some of these are:

- Drinking lots of water, juice or soft drinks.
- Taking rest breaks in a cool place.
- Wearing cool, protective clothing such as a shirt with collar and long sleeves and long trousers.
- Protecting the head – wear your hard hat.
- Applying SPF30+ sunscreen before exposure to sunlight as well as on over-cast days – noses, lips, ears, necks and backs of hands need extra protection. Sunscreen should be re-applied regularly. Provide communal dispensers.
- If possible, working in shaded areas in the high risk hours between 11 a.m. and 3 p.m.
- Not working near reflective surfaces such as water, cement, shiny metal or white painted sheds in strong sunlight.
- Protecting the eyes from glare.

26.30 What are the hazards and risks associated with welding fumes?

There are many different types of welding processes and the hazards and risks depend on what activity is being carried out.

Nearly all processes will produce metal fumes and gases. Some of the metal fumes will be:

- Aluminium
- Cadmium
- Chromium
- Copper
- Fluorides
- Iron
- Manganese
- Lead
- Molybdenum

- Nickel
- Tin
- Titanium
- Vanadium
- Zinc.

Some of the gas will be:

- Ozone
- Oxides of nitrogen
- Carbon monoxide.

The health hazards associated with welding fumes can be:

- Asthma
- Lung damage
- Emphysema
- Skin irritation
- Lung cancer
- Metal fume fever
- Respiratory irritation
- Siderosis – an accumulation of iron oxide in the lungs
- Lead poisoning
- Unconsciousness
- Death (especially from carbon monoxide).

The larger the exposure to the welding fumes and the higher the concentration of those fumes the greater the risk off ill-health.

26.31 What controls should be put in place on site to control the risks of exposure to welding fumes?

A wide range of controls can be implemented and where exposure to fumes is a site wide issue, the Principal Contractor shall take the collective measures necessary to protect the majority.

Controls could be:

- Control access to the work area – allow only those who need to be there.
- Locate the work away from openings, doors, windows (unless needed for ventilation but proper planning is needed).
- Provide good ventilation to the work area – approximately 5–10 air changes per hour.
- Provide an air speed of approximately 1 metre per second so as to clear fumes effectively.
- Provide local extraction equipment and ensure that it is working – regular check and monitor with a manometer.

- Discharge clean air outside of the building.
- Do not let fumes re-enter the building or discharge to other areas where people are working.
- Remove grease and any surface coatings from the materials to be welded.
- Ensure that the operative is in the correct position and that their head can be above any welding fumes generated.
- Extinguish the torch during rest periods so as to prevent nitrous fumes.
- Check the equipment to ensure that it is in good working order, all hoses, clips, etc. are in place.
- Ensure operatives wear the correct PPE.
- Keep records of checks.

27

Personal Protective Equipment

27.1 What are the legal requirements with regard to personal protective equipment?

The Personal Protective Equipment at Work Regulations 1992 applies in all work environments.

Employers have a duty to provide to the employees, *without charge*, all necessary personal protective equipment (PPE) which has been identified as necessary within the Risk Assessment.

PPE must be suitable and sufficient for use and appropriate for the risks it has been chosen to reduce. It must suit the worker for whom it is intended and afford adequate protection.

PPE must be comfortable to wear and fit properly to the user. If more than one type of PPE is to be used they must be compatible e.g. goggles and face mask must be suitable to be worn together.

Employers must ensure that employees have been given adequate information, instruction and training about how to wear PPE, why it has to be worn, its benefits, how to check it for defects, maintenance procedures, etc.

Suitable accommodation must be made available for the storage of PPE if it is impracticable for employees (and others) to store it on site.

Employers could be prosecuted for failing to supply appropriate PPE or for failing to maintain it in a suitable condition for use.

27.2 What are the responsibilities of the Site Agent with regard to issuing PPE?

The duty to issue PPEs rests with the employer or self-employed person and not with the Site Agent or Principal Contractor.

Where the Site Agent is the employer's on-site manager, then the employer has a duty to ensure that his *own* employees are correctly issued with PPE.

The Principal Contractor will determine the site rules for a construction site and in the rules it can stipulate the minimum requirement for PPE.

The Site Agent must ensure that his site rules are being followed – he has the duty to monitor and review safety on site.

27.3 Who determines whether a site is a "hard hat" site or not?

Generally, the Principal Contractor will determine whether the site will be hard hat site or not.

The applicable legislation is the Construction (Head Protection) Regulations 1989 and in particular Regulation 5 stipulates that the person having control of the site may make rules to ensure that others comply with the requirements of Regulation 4, which stipulates that an employer must ensure that suitable head protection is worn.

Rules about who, where and when head protection should be worn must be in writing and clearly displayed on the site.

Usually, all entry points, mess rooms and canteens display safety signs stating that "hard hats must be worn".

Wear there is "no foreseeable risk of head injury, other than by falling" hard hats do *not* need to be worn. In effect, the employer or Site Agent needs to complete a Risk Assessment which identifies the hazards and risks of working on the site.

If a Site Agent relaxes the hard hat rule, there should be a Risk Assessment which identifies that the risk of head injury is minimal.

If may be possible to relax the rule for hard hats for parts of the site, although this is not recommended as operatives may be confused as where and when to wear their hard hat.

Many fatalities and severe injuries are caused because employees and others do not wear protective hats. Brain damage is often irreversible and although someone may be physically very able, the damage done to the brain via a head injury could render the individual with a mentality of a child.

27.4 As a sub-contractor on site, can I decide that the area in which my employees are working does not have to be a hard hat area?

An as employer in control of your own site, you can decide when hard hats need to be worn but as a sub-contractor on a multi-occupied site you will have to follow the site rules and wear head protection.

Under the Construction (Design and Management) Regulations 2007, the Principal Contractor must make site rules with regard to the implementation and management of site safety. As a contractor, you have a duty to co-operate with the Principal Contractor and to follow the site rules.

There would be no harm in discussing the situation with the Site Agent with a view to seeking authority to relax the rules for head protection in your work area.

There may of course be other works being undertaken which you are not aware of and these could pose head injury hazards to your employees. It is the

Principal Contractor's duty to co-ordinate and manage the hazards and risks of other contractors on a multi-occupied site.

27.5 Is the Site Agent responsible for issuing PPE to site visitors?

It depends on the visitors! Usually, a Site Agent will provide PPE for the Client and Client's representatives as it would be unreasonable to expect the Client to provide the appropriate equipment. Hard hats are absolutely essential if there is a risk of falling objects or of hitting ones head on scaffold tubes, ceiling projections, etc.

It may not be reasonable for the Site Agent to provide PPE for:

- Architects
- Designers
- Quantity surveyors
- Building services consultants
- Project managers
- CDM Co-ordinators.

Unless it is *specialist* PPE which is site specific. All of the above will be employees of professional service firms or will be self-employed. The duty to provide PPE falls to the *employer*. They should provide their own!

27.6 What PPE would usually be expected on a main construction site?

The usual PPE comprises:

- Hard hats
- Safety boots/shoes
- Hi-vi vests/jackets
- Gloves
- Goggles/eye protectors.

It may be necessary to provide overalls and protective clothing as well.
It may also be necessary to provide hearing protection.

27.7 Should a Risk Assessment be completed in order to determine what PPE is required?

All work activities where there is a risk of injury to employees or others should be subjected to a full Risk Assessment.

Under the Management of Health and Safety at Work Regulations 1999, employers are responsible for completing Risk Assessments and for recording their significant findings.

Employers should therefore conduct Risk Assessments for any construction site-based activity and issue appropriate PPE.

The Site Agent should consider the hazards and risks in communal areas on the site and set down the findings in the Risk Assessment. From this, the site rules will be drawn up which stipulate what PPE is to be worn when and by whom.

The Site Agent may need to issue PPE to persons who are exposed to risk from hazardous substances from the activities of others.

28

Lifting Operations and Lifting Equipment

28.1 What is the legislation which governs the legal requirement for lifting and lifting equipment?

The Lifting Operations and Lifting Equipment Regulations (LOLER) 1998 contain all the legal requirements for the operating of safe lifting process and the use and maintenance of safe equipment.

The CDM Regulations do not specifically contain requirements for lifting but they require contractors, designers, clients and others to be aware of all legal requirements relating to the health and safety of a project and to comply with the legislation as necessary.

The Manual Handling Operations Regulations 1992 also contain requirements for lifting and the Principal Contractor should be aware of these in respect of lifting individual weights e.g. blocks, sacks, materials and so on.

Lifting equipment used at work is also subject to the general provisions of the Provision and Use of Work Equipment Regulations 1998.

28.2 What actually is lifting equipment?

Lifting equipment is defined as any equipment whose principal purpose is to lift or lower loads, including attachments used for anchoring, fixing and supporting it.

Lifting equipment can include:

- Cranes
- Fork lift trucks
- Jacks
- Mobile elevating work platforms
- Passenger lifts
- Vehicle tail lifts
- Ropes and pulleys
- Hoists

- Dumb waiters
- Vehicle inspection platforms.

  Lifting equipment also includes all lifting accessories such as:

- Chains
- Ropes
- Slings
- Shackles
- Eyebolts
- Harnesses
- Lanyards
- Running lines.

28.3 When do the Regulations apply?

The Regulations, commonly known as LOLER, apply to all work premises, including all construction sites. Hey apply to all work situations which are subject to the Health & Safety at Work Etc Act 1974.

The Principal Contractor must be familiar with the Regulations and should ensure that they are applied as appropriate to the construction site activities.

Should any contractor, or indeed any person, not be familiar with the Regulations, they should ask the competent person they have appointed under the Management of Health & Safety at Work Regulations 1999, or the CDM Co-ordinator appointed under CDM 2007, to advise them on how to apply the Regulations.

28.4 What are the principle requirements of LOLER 1998?

The Regulations require that the employer or person in control of the premises addresses the following:

- Suitability of lifting equipment
- Strength and stability
- Position and installation
- Marking of lift equipment
- Organization of lifting operations
- Examination and inspection
- Adequacy of equipment for lifting people.

Contracts Managers, Site Agents will need to ensure that either they, or the subcontractor providing and using the equipment, have addressed the key requirements.

28.5 What needs to be considered in respect of suitability of lifting equipment?

The following need to be considered:

- Ergonomic risks when using the equipment
- Suitability of material
- Safe means of access/egress
- Need to minimize risks of slips, trips and falls from any part of the lifting equipment
- Protection for operators especially in inclement weather
- Wind speeds, etc. and how high winds could affect either the equipment or the lifting operation.

There are many choices of lifting equipment available and for complex lifting tasks it will usually be advisable to seek competent advice. Some equipment will not be suitable for certain lifting tasks and choosing the wrong equipment could put people at risk of severe injury.

Ask the person choosing the equipment why they have chosen that type, have they considered all of the options, have they compiled Risk Assessments, etc.

28.6 What needs to be considered in respect of strength and stability of lifting equipment?

Equipment must have adequate strength for the proposed job/use, with an appropriate factor of safety against failure.

Equipment must also have adequate stability and not be prone to overturning.

Any equipment with pneumatic tyres must be checked to ensure the correct inflation of the tyres – under inflated tyres can cause equipment to topple over.

Spreader plates may be necessary or other types of stabilizing equipment. Conditions need to be checked to ensure that the stabilizers/spreaders can be fully extended and used correctly.

28.7 What are some of the key points to consider regarding the position and installation of lifting equipment?

Lifting equipment must be positioned or installed in such a way that the risk of a person being struck or a load moving in an uncontrolled manner is minimized.

If a load needs to be lifted over people this should be carefully planned and the position of the lifting equipment carefully considered so as to minimize the duration of the lift (also, the site agent should consider excluding all pedestrians and other vehicles from the area underneath the lifting span).

Equipment cannot be placed in any position where there is a likelihood that people could be crushed e.g. if it were to topple over, where would it fall and on to who?

Any path of travel of a load or the equipment itself should be protected by a suitable enclosure.

Lifting equipment in use should not come into contact with other lifting equipment e.g. proper planning and siting of multiple tower cranes on a site or the location of hoists not to be a risk of being entangled with netting and so on.

Access and egress points on to lifting equipment should be protected with suitable gates.

Interlocking devices should be fitted.

28.8 What is the safe working load of a lift or lifting equipment?

All lifts and lifting equipment must display the "safe working load (SWL)". This is the maximum weight that the equipment is designed to take safely.

Ignoring the SWL could cause equipment to be overloaded and plummet to the ground or lower levels.

Accidents happen because the SWL has been ignored.

Where there is a risk that the SWL will be ignored, the equipment should be fitted with interlocks or capacity limiters which prevent it being used in the overloaded state.

It is essential to check the SWL of slings, shackles, hoists, eye bolts, harnesses, lifting beams, etc.

If in doubt about the capacity of a lift or lifting equipment to take the load, seek advice from the manufacturer.

Where time is of the essence, reduce the load significantly and increase the number of times the loads have to be lifted – "little and often" may take longer but will reduce the risk of equipment collapse.

28.9 Organization of lifting operations seems to be critical for the safe operation of a construction site. What exactly does organization entail?

Lifting operations need to be:

- Properly planned
- Appropriately supervised
- Carried out in a safe manner.

The person planning the operation should have adequate practical and theoretical knowledge and experience of planning such operations. They *must* be competent.

It is good practice to write the Plan of Lifting Operations down as a record and aide memoire.

The plan must address:

- The risks identified
- The resources required
- The procedures to be followed
- Those involved and their responsibilities
- Any other lifts going on in the area and how they will interface.

Proper planning is a combination of:

- Initial planning
  - Is the equipment suitable
  - What is to be lifted
  - Weight, shape, centre of gravity, lifting points
  - Travel distance
  - Frequency
  - Weather conditions
  - Experience of operatives
- Appropriate consultation
  - Who will do the task
  - When does it need to be done
  - By when
  - Who might be affected
- Preparing method statements
  - How
  - Who
  - What
  - When
  - What could go wrong
- Contingency planning
  - Emergencies
  - Accidents
- Information, instruction and training
  - Evidence of training
  - Evidence of experience
  - Sharing of information re-site specific hazards e.g. overhead power lines, uneven ground
  - Itemizing preferred sequences of work
  - Detailing equipment to be used
  - Sharing the method statement.

28.10 Will the Principal Contractor have to supervise the lifting operation at all times in order to comply with the LOLER 1998?

No. But all lifting operations will need to be supervised by a competent person. This may not necessarily be the Principal Contractor but could be the specialist contractor engaged to carry out the lifting operation.

The Principal Contractor must be aware of the lifting activity to be undertaken and should have a copy of the written lifting plan.

The Principal Contractor is not expected to be an expert in lifting operations but is expected to be competent enough to assess whether due regard has been paid to health and safety and compliance with LOLER 1998.

The Principal Contractor must ensure that someone competent is supervising the lifting operations. It would be good practice to record this on the written plan.

The Principal Contractor should review the operation once it is underway and ensure that the work plan/method statement is being followed. This will be especially necessary where the Principal Contractor is responsible for the safety of others and the lifting operation involves loads being swung over people's heads.

The Principal Contractor will have the authority to stop the lifting operation – or the use of lifting equipment if it is deemed unsafe, or if, for instance, the weather conditions have deteriorated.

28.11 What is required regarding inspection and testing of lifting equipment?

The Regulations require that lifting equipment is thoroughly examined by a competent person to ensure that it is suitable for use.

Thorough examination should be carried out at regular intervals and at least:

- Every 6 months if it is used for lifting people or is an accessory used for lifting people
- At least every twelve months if it is used for lifting goods.

Thorough examination may be required more frequently than the above if specified by the competent person.

Records of all thorough examinations must be kept.

The Principal Contractor should check the records of thorough examination of all lifting equipment being brought onto site.

Regular inspection of lifting equipment is not the same as thorough examination and can be carried out much more frequently.

All users of lifting equipment should be trained to carry out visual checks and routine rests of their equipment before using it.

Lanyards, harnesses, ropes, slings, etc. should all be regularly checked for:

- Fraying strands
- Cut surfaces
- Contamination with chemicals
- Defective karibiners, etc.
- Over stretching
- Faulty mechanisms.

Any equipment found to be faulty must be put out of use, labelled accordingly and returned to the supplier for disposal/repair.

The records of lifting equipment can be disposed of as soon as the equipment ceases to be used or operated.

Any reports for lifting accessories e.g. eyebolts, harnesses, slings and so on must be kept for a minimum of 2 years after the report was made.

Reports on passenger lifts or permanent lifting equipment within the building must be kept as long as the equipment is used and until the next formal report is issued. Reports should be kept on the premises to where the equipment is installed.

It is good practice to keep all records of inspection and testing as this will show a "due diligence" defence should things go wrong. If you can demonstrate to the Court that you were aware of your duties and kept records and that any accident was not through any negligence of your own, you may well be able to show that you had done everything reasonably practicable to avoid the commission of an offence.

Lifting Equipment/Operations Checklist			
		Yes	No
1	Do you carry out any lifting operations or have any lifting equipment on site which is subject to the Lifting Operations & Lifting Equipment Regulations 1998?		
2	Do you ensure that equipment is: • Suitable • Of adequate strength • Stable • Suitable for lifting people and meets any specific requirements for this task		
3	Is all lifting equipment and lifting accessories marked with the SWL?		

	If not individually marked, is there adequate signage close to the equipment to advise operatives of the SWL?		
4	Are all lifting operations properly planned, with a written plan available?		
5	Are all lifting operations properly supervised by competent persons?		
6	Have you checked all the thorough examination records for the lifting equipment? Are they the correct ones for the equipment?		
7	Are all operatives trained, instructed and informed about the lifting operations, the site-specific hazards, any acceptable risks, etc.?		
8	Are operatives trained to carry out visual inspections and to carry out function tests of the equipment they are to use? Have spot checks been undertaken?		
9	Have to instigated a system of immediate reporting of defective equipment and subsequent remedial actions?		
10	Have you kept records for the appropriate time and will relevant information be passed to the building user i.e. permanent passenger lift thorough examination reports, eyebolt reports when used for fall arrest systems?		

Case Studies

Incident

Vehicle crane operator killed after his vehicle's lifting equipment came into contact with overhead power lines as he delivered materials.

What would have prevented the incident?

- Proper Planning of the job
- Identification of overhead lines
- Site-specific risk assessment
- Review of the equipment to be used
- Level of competency of the operator
- On-site supervision.

Incident

A man sustained life threatening injuries after being thrown from the platform of a mobile elevating work platform as he worked in the street to repair the fabric of residential buildings. It is alleged that a passing vehicle struck the platforms support as it passed.

What would have prevented the incident?

- Proper planning of the job
- Choice of appropriate lifting equipment which may have had a different type of stabiliser
- Appropriate barriers and control of passing traffic
- Operative should have been wearing fall arrest equipment – harness and lanyard
- Adequate supervision.

Incident

A man died in a vehicle picked up and crushed by a hydraulic grab machine at a car breakers yard. The man was a member of the public visiting the site and it is thought he might have entered the vehicle just as it was being lifted.

What would have prevented the incident?

- Proper supervision and control of all work procedures
- Strict controls on public access to areas
- Prohibition of any lifting activity in the area where people are present
- Better grab operator training
- Giving visitors Hi-Vi vests/jackets
- Better planning of operations
- Risk assessments and method statements.

29

Manual Handling

29.1 What are the main hazards and risks associated with manual handling activities?

Manual handling activities are responsible for millions of lost time accidents and injuries and over 50% of all over 3 day accidents involve manual handling of some sort.

"Bad backs" are common and are not always caused by lifting exceptionally heavy weights. Often it is an accumulation of inappropriate lifting techniques, bad posture and poor systems of work. Sometimes, bending down to pick up a hand held tool will be sufficient to cause a muscular spasm, slipped disc or other injury.

The main hazards are:

- Lifting too heavy a weight
- Lifting awkward weights
- Tripping whilst carrying
- Pushing, pulling and shoving loads
- Repetitive lifting
- Repetitive actions
- Dropping weights/loads.

The hazards, i.e. the activities with the potential to cause harm, will often result in the following injuries:

- Slipped discs
- Pulled muscles
- Strained muscles
- Torn muscles
- Torn ligaments
- Cracked bones e.g. ribs
- Repetitive strain injuries
- Impact injuries
- Hernias.

29.2 What are the legal requirements regarding manual handling activities?

The Manual Handling Operations Regulations 1992 set down the legal duties surrounding manual handling activities at work.

Employers are required to complete Risk Assessments for all manual handling activities carried on at work, and the significant findings must be recorded in writing.

The responsibility under the Regulations for the employer is to eliminate or reduce the amount of manual handling undertaken by employees.

The risk of injury must be reduced to the lowest level possible.

29.3 What are some of the practical steps which can be taken to reduce manual handling on construction sites?

The first duty to reduce manual handling activities on construction sites sits with designers under the CDM Regulations 2007.

Designers have a duty to design out hazards associated with their designs both in relation to the future use of the building and to how it will be constructed.

The CDM Co-ordinator should assist in ensuring that the Designers have considered their duties and provided Design Risk Assessments for the Principal Contractor.

Designers should consider:

- The weight of materials they specify
- The size of the materials they specify
- The shape of materials they specify
- The type of materials they specify
- The sequence of construction.

The CDM Co-ordinator should encourage Designers to reduce the likelihood of injury from manual handling activities. Consideration should also be given as to how materials will be delivered and moved around site e.g. carrying $3\,m \times 2\,m$ boards up 10 flights of stairs may be excessive manual handling and the Designer should have considered smaller boards or hoist access into the building.

Materials should be available in the lightest weights possible. Bags, blocks, etc. should weigh 25 kilogram or less.

Once the Principal Contractor is in receipt of any Design Risk Assessments and the Pre-Construction Health and Safety Pack (as this may outline the philosophy of the design in respect of manual handling), the Site Agent should carry out a Risk Assessment of the materials needed on site, the sizes specified in the design, access and movement of materials around the site and where they need to be used. Any concerns should be raised if appropriate, with the CDM Co-ordinator who should then liaise with the Design Team and seek solutions.

Manual handling must be eliminated where possible and in any event reduced to acceptable levels. Small improvements in manual handling will have a significant long term impact and must not be ignored.

Mechanical aids, lifts, hoists, trolleys, etc. eliminate most of the manual handling activity and the Risk Assessment should consider these options.

Plan what equipment is needed throughout the site.

Avoid double handling of materials – it increases the risk of injury.

Reduce the height to which materials need to be lifted by positioning loads at height by mechanical aids scissor lifts and scissor tables could assist.

Reduce the distance materials need to be carried – consider the delivery plan and where materials will be stored, where they will be needed, how often they need to be replenished, etc.

Encourage the sharing or lifting of loads which are too heavy for one person or which are of an awkward shape. Two people lifts will be (usually) safe than one person lifts.

29.4 What are the key steps in a manual handling risk assessment?

There are four key elements to undertaking a manual handling Risk Assessment, namely:

- Consider the task
- Consider the load
- Consider the working environment
- Consider the individual capability.

29.4.1 The tasks

Do they include:

- Holding or manipulating loads at a distance from the trunk
- Unsatisfactory bodily movement or posture
- Twisting the trunk
- Stooping
- Reaching upwards
- Excessive movement of loads, especially the following:
  - Excessive lifting or lowering distances
  - Excessive carrying distances
- Excessive pushing or pulling of loads
- Risk of sudden movement of loads
- Frequent or prolonged physical effort

- Insufficient rest or recovery periods
- A rate of work imposed by a process.

29.4.2 The loads

Are they:

- Heavy
- Bulky or unwieldy
- Difficult to grasp
- Unstable or with contents likely to shift
- Containing sharp edges, protruding nails, screws, etc.
- Hot from process.

29.4.3 The working environment

Is there:

- Space constraints due to small work areas, restricted heights, etc. which will prevent good lifting postures
- Uneven, slippery, defective, unstable floors
- Changes in floor level either via steps or ramps
- Extreme temperature changes which could affect physical exertion
- Ventilation/air changes which could raise dust, cause gusts of wind, etc.
- Overcrowding of work spaces
- Physical obstructions
- Poor lighting.

29.4.4 Individual capacity

Does the job:

- Require unusual strength
- Require special knowledge of the load
- Create any other hazards to the individual carrying the load.

The information collected in the stages above will form the basis of assessing the hazards and risks from the manual handling task to be undertaken.

The likelihood of injury will range from high to low depending on the inter-relationship of the above factors.

Control measures will be identified to reduce the risk of injury e.g.:

- Remove obstructions
- Increase lighting
- Reduce load weight
- Use mechanical aid
- Provide training on lifting techniques.

When the control measures are implemented the risk of injury will reduce.

The Risk Assessment will need to be reviewed regularly for the same manual handling process.

29.5 Can manual handling be done safely?

There are several key rules for safe manual handling. Information should be freely available on safe lifting techniques in the mess room, site office, etc. so that operatives can be reminded of the correct way to lift objects.

The fundamentals of safe lifting are:

- Take a secure grip
- Use the proper feet position – feet apart with the leading foot pointing in the direction you intend to go
- Adopt a position with bent knees but a *straight* back
- Keep arms close to the body
- Keep head and chin tucked in
- Keep the load close to the body
- Use body weight where possible
- Push up for the lift using the thigh muscles.

Operatives should also consider the environment they are to lift in and pre-plan any interim rest positions, how to change direction, etc. The Principal Contractor should oversee manual handling activities which are site wide.

29.6 What topics should be covered in a manual handling tool box talk?

An effective tool box talk agenda will cover:

- What checks to carry out before starting manual handling
- How to judge your capability
- Environmental conditions
- Wearing of personal protective equipment e.g. gloves, safety boots
- Carrying out a trial lift first
- Getting help – its not seen as being a wimp
- Good handling techniques
- Reducing the weights of objects
- Using mechanical aids
- Checks to carry out during the manual handling tasks e.g. obstructions in the travel route and so on.

Manual handling improvements: Evaluation

There are many ways to improve manual handling on a construction site and when carrying out refurbishment works.

But sometimes what seems to be an improvement in one area creates additional hazards or risks in another.

The following questions might help to ensure that one hazard is not being substituted for another.

Will the improvement:

- Reduce or eliminate most or all of the identified risk factors
- Add any new risks that have not been previously identified
- Be affordable i.e. is there a simpler, less expensive option which would be equally effective
- Affect productivity of effectiveness
- Affect product or service quality
- Provide a temporary or permanent fix
- Be accepted by site operatives
- Be fully implemented, including any training, within a reasonable time
- Affect any specific trades more than others thus affecting work rate, etc.

Will the equipment:

- Reach far enough to cover the work area
- Handle the weight and shape of the product
- Re-orient the load as needed
- Be easy to load/unload
- Require much force or energy to push it, steer it or stop at the destination given the typical ground conditions of a construction site
- Be heavy or large and create its own manual handling issues
- Handle the load in a safe and controlled manner
- Hold the load securely and well balanced
- Allow too much movement from any chains or cables if used
- Allow adequate field of vision for the operator
- Slow operatives down too much thus causing them to take shortcuts and create greater risks
- Interface with existing equipment and structures
- Obstruct the movement of people, materials or equipment around the rest of the site
- Need an additional power supply beyond the capacity of the system already in place.

Manual Handling Assessment Form Manual Handling Checklist					
Activity:					
Questions to consider: (If the answer to a question is "yes" place a tick against it and then consider the level of risk)	**Level of risk:** (Tick as appropriate)				**Possible remedial action:**
	Yes	Low	Med	High	
The tasks – do they involve: • holding loads away from body? • twisting? • stooping? • reaching upwards? • large vertical movement? • long carrying distances? • strenuous pushing or pulling? • unpredictable movement of loads? • repetitive handling? • insufficient rest or recovery? • a work rate imposed by a process?					
The loads – are they: • heavy? • bulky/unwieldy? • difficult to grasp? • unstable/ unpredictable? • intrinsically harmful (e.g. sharp/hot?)					

The working environment – are there: • constraints on posture? • poor floors? • variations in levels? • hot/cold/humid conditions? • strong air movements? • Poor lighting conditions?					
Individual capability – does the job: • require unusual capability? • hazard those with a health problem? • hazard those who are pregnant? • call for special information/training					
Other factors – Is movement or posture hindered by clothing or personal protective equipment?					

Deciding the level of risk will inevitably call for judgement, use the information contained within the policy document to assist you.

30

Fire Safety

30.1 What causes a fire to start?

A fire needs the following to start:

(1)   Fuel
(2)   Ignition
(3)   Oxygen.

The above three component parts are often referred to as the "Triangle of Fire".

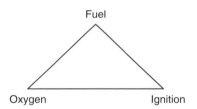

If one or more of the component parts of a fire are eliminated a fire will not start or continue.

As the elements of fire are not successfully managed on all construction sites, the industry suffers approximately 4,000 fires a year, costing millions of pounds in destroyed buildings, materials, plant and equipment, and delayed projects.

Other insurance cover for buildings and construction work is difficult to obtain without a proper fire safety management plan in place.

30.2 What is the Regulatory Reform (Fire Safety) Order 2005?

The Order contains the legal requirements for managing fire safety in premises.

It repeals virtually all existing legislation by consolidating previous requirements into one statutory instrument.

30.3 What type of premises does the Order apply to?

The Order applies to virtually all premises and covers nearly every type of building, including open spaces.

The order applies to:

- Offices
- Shops
- Factories
- Care homes
- Hospitals
- Churches, places of worship
- Village halls, community halls
- Schools and colleges, universities
- Pubs, clubs, restaurants
- Sports centres
- Tents and marquees
- Hotels, hostels
- Warehouses
- Shopping centres
- Construction sites.

The Order does not apply to private dwellings including individual flats in a block or house.

(N.B.: Fire safety requirements for houses in multiple occupation are contained in Housing Act legislation).

30.4 Who is legally responsible for complying with the Regulatory Reform (Fire Safety) Order 2005?

The Order refers to a "responsible person" as being the person with responsibility for complying with the Order.

In the workplace, the responsible person will be the employer or any other person who may have control of any part of the premises e.g. owner or occupier.

In all other premises the person or persons in control of the premises will be responsible for complying with the Order.

30.5 What happens if there is more than one person in control of premises?

Where there is more than one responsible person in any type of premises e.g. multi-occupied complex, all must take reasonable steps to co-operate with one another.

If there are managing agents responsible for common parts then they will be the responsible person for those areas.

The requirements of the Order are in fact imposed on *any* person who has, to any extent, control of premises so far as the requirements of the Order relate to matters within his control.

30.6 What are the main requirements or rules under the Order?

The responsible person must:

- Carry out a Fire Risk Assessment identifying any possible dangers and risks of fire.
- Consider who may be especially at risk.
- Reduce the risk of fire within the premises as far as is reasonably practicable.
- Provide general fire precautions to deal with any residual fire risk.
- Implement special measures to control the risks from flammable materials, explosive materials or other hazardous substances.
- Create an emergency plan for fire safety.
- Record significant findings in writing.
- Review and monitor fire safety arrangements.

In addition, the responsible person must ensure everybody who could be affected by their business undertaking receives information or training and instruction.

The responsible person must also prepare an emergency plan for fire safety arrangements.

Any equipment, facilities e.g. fire fighting, lifts, provided for the use and safety of fire fighters must be adequately maintained.

30.7 Does the duty to ensure fire safety of persons relate only to employees?

The Order refers to all "relevant persons".

This includes everyone – employees, public, customers, contractors, visitors, students, patients, etc.

The responsible person must consider the fire safety needs of those at special risk e.g. disabled people, young people and people not familiar with the building.

The Regulatory Reform (Fire Safety) Order 2005 introduces clear responsibility for the "responsible person" to consider the needs of the public in respect of fire safety and not just employees.

30.8 What are the main requirements of the Order?

The main requirement of the Order is for the responsible person to carry out a Fire Risk Assessment.

The Fire Risk Assessment must focus on the safety in case of fire of all "relevant persons".

How to undertake a Fire Risk Assessment is covered in detail in Chapter 3 of this Guide.

In addition to a Fire Risk Assessment, the responsible person must:

- Appoint one or more competent persons to assist with fire safety.
- Provide employees with clear and relevant information on the risks to them, identified by the Fire Risk Assessment, about measures taken to prevent fires and generally, how they will be protected in the event that a fire occurs.
- Consult with employees about fire safety matters.
- Take special steps regarding the employment of young people and inform their parents about the fire measures that have been taken.
- Inform all persons who are not employees about fire safety measures in the premises e.g. contractors and temporary workers.
- Co-operate with other responsible persons and co-ordinate fire safety matters in multi-occupied buildings.
- Inform the employer of any temporary workers from an outside organization with clear and relevant information about how the safety of their employees will be protected in the event of fire.
- Consider any special precautions needed for managing fire safety from any dangerous, flammable or hazardous substances, including explosive.
- Establish a suitable means for contacting the emergency services and providing them with relevant information about dangerous substances.
- Provide all relevant persons, where reasonable, with suitable information, instruction and training on fire safety.
- Provide and maintain suitable fire fighting equipment, fire detection equipment, fire warning systems, emergency lighting, etc. as is deemed necessary in the premises by the Fire Risk Assessment.
- Advise employees that they have a duty to co-operate with their employer so as to enable them to comply with their statutory duties.

30.9 Can anyone be appointed as the competent person?

Yes, provided they have enough knowledge and experience to understand the basic principles of fire safety and to be able to implement the measures contained in the Regulatory Reform (Fire Safety) Order 2005.

There are no formal qualifications to become a competent person, but as the onus is on the responsible person to appoint someone – or more than one person if necessary – to provide safety assistance, the responsible person will need to be able to demonstrate that the person so appointed has suitable credentials to take on the role.

The type of person who could be classed as competent person could be:

- Fire Safety Consultants
- Fire Engineers
- Health and Safety Managers/Officers – provided that have fire safety training
- Employers – if they have attended basic fire safety training
- Unit/Department Managers – with fire safety training
- Health and Safety Consultants.

The competent person is there to help the responsible person to fulfil their statutory duties.

30.10 What are the principles of prevention in respect of fire safety?

The RRO requires all responsible persons to undertake Fire Risk Assessments of their premises, operations and any other business activity which could expose persons to risks to their safety.

If the Risk Assessment indicates that preventative and protective measures are required then those measures must be in accordance with the requirements of Schedule 1, Part 3 of the RRO.

The principles of prevention are:

(a) Avoiding risks.
(b) Evaluating the risks which cannot be avoided.
(c) Combating risks at source.
(d) Adapting to technical progress.
(e) Replacing the dangerous with the non-dangerous or less dangerous.
(f) Developing a coherent overall protection policy which covers technology, organization of work and the influence of factors relating to the working environment.
(g) Giving collective protective measures priority over individual protective measures.
(h) Giving appropriate instructions to employees.

The principles of prevention listed above are more detailed than in other health and safety legislation.

30.11 The Fire Officer has indicated that my Fire Risk Assessment is not "suitable and sufficient". What should I do about it?

If possible, discuss it with the Fire Officer. Sometimes they do not like the format of a Risk Assessment rather than the content!

The Fire Officer may be concerned that you have not considered all aspects of fire safety and that your Risk Assessment is missing a vital element. Generally, the Fire Officer should tell you why they are not satisfied with your Risk Assessment.

If you feel that you have covered everything and that the Fire Officer is being over-zealous, then ask to discuss the matter with a Senior Officer.

Be prepared and review your Risk Assessment together with your competent person. Or seek professional advice.

A Fire Risk Assessments must address the major fire safety hazards associated with your workplace – it does not have to be perfect.

30.12 Where can I obtain further information on complying with the Regulatory Reform (Fire Safety) Order 2005?

All Fire Authorities should be able to give you information and advice on complying with fire safety requirements.

However, in order to address the needs of the business community in understanding the duties placed on them by RRO, the Government have published a series of Fire Safety Risk Assessment guides. These guides set out everything an employer – or responsible person – will need to do to comply with the law and ensure fire safety standards are maintained.

The guides cover:

- Offices and shops
- Factories and warehouses
- Sleeping accommodation
- Small and medium places of assembly
- Large places of assembly
- Theatres and cinemas
- Educational premises
- Residential care premises
- Outdoor events
- Healthcare premises
- Transport premises and facilities.

All the guides are available for free download on the government website: www.firesafetyguides.communities.gov.uk.

30.13 What is the legal requirement for a fire risk assessment?

The Regulatory Reform (Fire Safety) Order 2005 requires the "responsible person" to make a suitable and sufficient assessment of the risks to which relevant persons are exposed in respect of fire.

Any premises in which persons are employed or to which others have access must be subject to a fire risk assessment.

The principles of risk assessment to be followed are the same as those as listed in the Management of Health and Safety at Work Regulations 1999.

Where there are five or more employees, the significant findings of the risk assessment must be recorded in writing.

30.14 What actually is a fire risk assessment?

A fire risk assessment is in effect an audit of your workplace and work activities in order to establish how likely a fire is to start, where it would be, how severe it might be, who it would affect and how people would get out of the building in an emergency.

A fire risk assessment should be concerned with *life safety* and not with matters which are really fire engineering matters.

A fire risk assessment is a structured way of looking at the hazards and risks associated with fire and the products of fire e.g. smoke.

Like all risk assessments, a fire risk assessment follows *five key steps*, namely:

(i)   Identify the hazards
(ii) Identify the people and the location of people at significant risk from a fire
(iii) Evaluate the risks
(iv) Record findings and actions taken
(v)   Keep assessment under review.

So, a fire risk assessment is a record that shows you have assessed the likelihood of a fire occurring in your workplace, identified who could be harmed and how, decided on what steps you need to take to reduce the likelihood of a fire (and therefore its harmful consequences) occurring. You have recorded all these findings regarding your pub into a particular format, called a risk assessment.

Definitions

- *Risk assessment*
  - The overall process of estimating the magnitude of risk and deciding whether or not the risk is tolerable or acceptable.

- *Risk*
  - The combination of the likelihood and consequence of a specified hazardous event occurring.
- *Hazard*
  - A source or a situation with a potential to harm in terms of human injury or ill-health, damage to property, damage to the environment or a combination of these.
- *Hazard identification*
  - The process of recognizing that a hazard exists and defining its characteristic.

30.15 What are the *five* steps to a fire risk assessment?

30.15.1 Step 1: Identify the hazards

Sources of ignition

You can identify the sources of ignition in your premises by looking for possible sources of heat that could get hot enough to ignite the material in the vicinity.

Such sources of heat/ignition could be:

- Smoker's materials
- Naked flames e.g. candles, fires, blow lamps and so on.
- Electrical, gas or oil fired heaters
- Hot work processes e.g. welding, gas cutting
- Cooking, especially frying
- Faulty or misused electrical appliances including plugs, extension leads
- Lighting equipment, especially halogen lamps
- Hot surfaces and obstructions of ventilation grill e.g. radiators
- Poorly maintained equipment that causes friction or sparks
- Static electricity
- Arson.

Look out for any evidence that things or items have suffered scorching or overheating e.g. blackened plugs and sockets, burn marks, cigarette burns, scorch marks and so on.

Check each area of the premises systematically:

- Customer area/public areas/reception
- Work areas, offices
- Staff kitchen, staff rooms
- Store rooms, cleaners stores
- Plant rooms, motor rooms

- Refuse areas
- External areas.

Sources of fuel

Anything (generally) that burns is fuel for a fire. Fuel can also be invisible in the form of vapours, fumes, etc. given off from other less flammable materials.

Look for anything in the premises that is in sufficient quantity to burn reasonably easily, or to cause a fire to spread to more easily accessible fuels.

Fuels to look out for are:

- Wood, paper, cardboard
- Flammable chemicals e.g. cleaning materials
- Flammable liquids e.g. cleaning substances, liquid petroleum gas
- Flammable liquids and solvents e.g. white spirit, petrol, methylated spirit
- Paints, varnishes, thinners, etc.
- Furniture, fixtures and fittings
- Textiles
- Ceiling tiles and polystyrene products
- Waste materials, general rubbish
- Gases.

Consider also the construction of the premises – are there any materials used which would burn more easily than other types. Hardboard, chipboard and blockboard burn more easily than plasterboard.

Identifying sources of oxygen

Mostly oxygen is all round us in the air that we breathe. Sometimes, other sources of oxygen are present that accelerate the speed at which a fire ignites e.g. oxygen cylinders for welding.

The more turbulent the air the more likely the spread of fire will be e.g. opening doors brings a "whoosh" of air into a room and the fire is fanned and intensifies. Mechanical ventilation also moves air around in greater volumes and more quickly.

Do not forget that whilst ventilation systems move oxygen around at greater volumes, it also will transport smoke and toxic fumes around the building.

30.15.2 Step 2: Identify who could be harmed

You need to identify who will be at risk from a fire and where they will be when a fire starts. The law requires you to ensure the safety of your staff and others e.g. customers. Would anyone be affected by a fire in an area that is isolated? Could everyone respond to an alarm, or evacuate?

Will you have people with disabilities in the premises e.g. wheelchairs, visually or hearing impaired? Will they be at any greater risk of being harmed by a fire than other people?

Will contractors working in plant rooms, on the roof, etc. be adversely affected by a fire? Could they be trapped or not hear alarms?

Who might be affected by smoke travelling through the building? Smoke often contains toxic fumes.

30.15.3 Step 3: Evaluate the risks arising from the hazards

What will happen if there is a fire? Does it matter whether it is a minor or major fire? Remember small fires can grow rapidly to infernos.

A fire is often likely to start because:

(i) People get careless with cigarettes and matches
(ii) People purposely set light to things
(iii) Cooking canopies catch fire due to grease build up
(iv) People put combustible material near flames/ignition sources
(v) Equipment is faulty because it is not maintained
(vi) Electrical sockets are overloaded.

Will people die in a fire from:

(i) Flames
(ii) Heat
(iii) Smoke
(iv) Toxic fumes.

Will people get trapped in the building?

Will people know that there is a fire and will they be able to get out?

Step 3 of the risk assessment is about looking at what *control measures* you have in place to help control the risk or reduce the risk of harm from a fire.

Remember – fire safety is about *life safety*. Get people out fast and protect their lives. Property is always replaceable.

You will need to record on your fire risk assessment the fire precautions you have in place i.e.:

(i) What emergency exits do you have and are they adequate and in the correct place?
(ii) Are they easily identified, unobstructed?
(iii) Is there fire fighting equipment?
(iv) How is the fire alarm raised?
(v) Where do people go when they leave the building – an assembly point?
(vi) Are the signs for fire safety adequate?
(vii) Who will check the building and take charge of an incident i.e. do you have a Fire Warden appointed?

(viii) Are fire doors kept closed?

(ix) Are ignition sources controlled and fuel sources managed?

(x) Do you have procedures to manage contractors? (Remember Windsor Castle went up in flames because a contractor used a blow torch near the curtains!)

Of all your fire safety precautions for the premises, is there anything more that you need to do?

Are staff trained in what to do in an emergency? Can they use fire extinguishers? Do you have fire drills? Is equipment serviced and checked e.g. emergency lights, fire alarm bells and so on.

30.15.4 Step 4: Record findings and action taken

Complete a fire risk assessment form and keep it safe.

Make sure that you share the information with staff.

If contractors come to site, make sure that you discuss *their* fire safety plans with them and that you tell them what your fire precaution procedures are.

30.15.5 Step 5: Keep assessment under review

A fire risk assessment needs to be reviewed regularly – about every 6 months or so and whenever something has changed, including layout, new employees, new procedures, new legislation, increased stock, etc.

30.16 Is there any guidance on assessing the risk rating of premises in respect of fire safety?

When completing fire risk assessments it is sensible to categorize *residual risk* for the premises into a risk rating category – normally referred to a high, medium or low.

In terms of fire risk rating it is usual to refer to medium risk as "normal".

The government's publication "Fire Safety – An Employers Guide" gives some guidance on how to fire risk rate premises. Also, the new Fire Safety Guides published to support the Regulatory Reform (Fire Safety) Order 2005 also gives some advice on risk rating premises.

30.16.1 High-risk premises

- Any premises where highly flammable or explosive substances are stored or used (other than very small quantities).

- Any premises where the structural elements present are unsatisfactory in respect of fire safety:
  - Lack of fire-resisting separation.
  - Vertical or horizontal openings through which fire, heat and smoke can spread.
  - Long and complex escape routes created by extensive sub-division of floors by partitions, etc.
  - Complex escape routes created by the positioning of shop unit displays, machinery, etc.
  - Large areas of smoke or flame producing furnishings and surface materials especially on walls and ceilings.
- Permanent or temporary work activities which have the potential for fires to start and spread e.g.:
  - Workshops using highly flammable materials and substances.
  - Paint spraying.
  - Activities using naked flames e.g. blow torches, welding.
  - Large kitchens in work canteens, restaurants.
  - Refuse chambers and waste disposal areas.
  - Areas containing foam or foam plastic upholstery and furniture.
- Where there is significant risk to life in case of fire:
  - Sleeping accommodation provided for staff, guests, visitors, etc. in significant numbers.
  - Treatment of care where occupants have to rely on others to help them
  - High proportions of elderly or infirm.
  - Large numbers of people with disabilities.
  - Where people work in remote areas e.g. plant rooms, roof areas and so on.
  - Large numbers of people resort to the premises relative to its size e.g. sales at retail shops.
  - Large numbers of people resorting to the premises where the number of people available to assist is limited e.g. entertainment events, banquets and so on.

30.16.2 Normal-risk premises

- Where an outbreak of fire is likely to remain contained to localized areas or is likely to spread only slowly, allowing people to escape to a place of safety.
- Where the number of people in the premises is small and they are likely to escape via well-defined means of escape to a place of safety without assistance.
- Where the premises has an automatic warning system or an effective automatic fire fighting/fire extinguishing/fire suppression system which may reduce the risk categorization from high.

30.16.3 Low-risk premises

- Where there is minimal risk to peoples' lives and where the risk of fire occurring is low or the potential for fire, heat or smoke spreading is negligible.

30.17 Who must I consider when preparing my Risk Assessment?

Responsible persons must consider the following people as being at risk in the event of a fire:

- Employees and those on temporary or agency contracts
- Employees whose mobility, sight or hearing might be impaired
- Employees with learning difficulties or mental illness
- Other persons in the premises if the premises are multi-occupied
- Anyone occupying remote areas of the premises
- Visitors and members of the public, including contractors, etc.
- Anyone who may sleep on the premises
- Anyone with any special needs or disabilities.

30.18 Does a fire risk assessment have to consider members of the public?

A fire risk assessment must be carried out by the responsible person and must consider the risks to the safety of "*relevant persons*" i.e. all persons who are, or could be, lawfully on the premises. This will include members of the public.

The Regulatory Reform (Fire Safety) Order 2005 has made the inclusion of all persons, including the public, a legal requirement when completing risk assessments.

30.19 Who can carry out a fire risk assessment?

The Regulatory Reform (Fire Safety) Order 2005 states that the person who carries out a fire risk assessment shall be *competent* to do so. They do not necessarily have to have had formal training.

Competency is not defined specifically in the Order but is generally taken to mean having a level of knowledge and experience which is relevant to the task in hand.

A Fire Risk Assessment is a logical, practical review of the likelihood of a fire starting in the premises and the consequences of such a fire. Someone who has good knowledge of the work activities and the layout of the building, together with some knowledge of what causes a fire, would be best placed to carry out a fire risk assessment.

30.20 What fire hazards need to be considered?

Consider any significant fire hazards in the room or area under review.

- Combustible materials e.g. large quantities of paper, combustible fabrics, plastics.
- Flammable substances e.g. paints, thinners, chemicals, flammable gases, aerosol cans.
- Ignition sources e.g. naked flames, sparks, portable heaters, smoking materials, hot works equipment.

Do not forget to consider materials which might smoulder and produce quantities of smoke. Also, consider anything which might be able to give off toxic fumes.

Consider as well the type of insulation involved or used in cavities, roof voids, etc. Combustible material may not always be visible e.g. hidden cables in wall cavities.

30.21 What structural features are important to consider when carrying out a fire risk assessment?

Fire, smoke, heat and fumes can travel rapidly through a building if it is not restricted by fire protection and compartmentation.

Any part of a building which has open areas, open staircases, etc. will be more vulnerable to the risk of fire should one start.

Openings in walls, large voids above ceilings and below floors allow a fire to spread rapidly. Large voids also usually contribute extra ventilation thereby adding more oxygen to the fire.

A method of fire prevention is to use fire resistant materials and to design buildings so that fire will not travel from one are to another.

Any opportunity for a fire to spread through the building must be noted on the fire risk assessment.

30.22 What are some of the factors to consider when assessing existing control measures for managing fire safety?

Many premises and employers already have some level of fire safety management in place and the Regulatory Reform (Fire Safety) Order 2005 is not intended to add an especially heavy burden on to employers and others as responsible persons.

Existing control measures must be reviewed and the following are examples of what to look for:

- The likely spread of fire
- The likelihood of fire starting
- The number of occupiers of the area
- The use and activity undertaken
- The time available for escape
- The means of escape
- The clarity of the escape
- Effectiveness of signage
- How the fire alarm is raised
- Can it be heard by everyone
- Travel distances
- Number and widths of exits
- Condition of corridors
- Storage and obstructions
- Inner rooms and dead ends
- Type and access to staircases
- Openings, voids, etc. within the building
- Type of fire doors
- Use of panic bolts
- Unobstructed fire doors
- Intumescent strips
- Well fitting fire doors
- Propped open fire doors
- Type of fire alarm
- Location, number and condition of fire extinguishers
- Display of fire safety notices
- Emergency lighting
- Maintenance and testing of fire alarm break glass points
- Installation of sprinkles
- Location and condition of smoke detectors
- Use of heat detectors
- Adequate lighting in an evacuation
- Training of employees
- Practised fire drills
- General housekeeping
- Management of contractors
- Use of hot works permits
- Control of smoking
- Fire safety checks
- Provision for managing the safety of people with disabilities
- Special conditions e.g. storage of flammable substances
- Storage of combustible materials near a heat source.

The best fire risk assessments are "site specific" – review and inspect your *own* workplace. Example templates for Fire Risk Assessments follow.

30.23 What is meant by the term means of escape?

The term "means of escape" in fire safety refers to the routes which are provided in a building for occupants to leave the building in an emergency, quickly and safely without unnecessary assistance.

Means of escape are usually alternative exits from a building which have been designated to be accessed quickly from any occupied area of the building.

Any means of escape should lead to a place of safety, they should be easy to identify and safe to use.

Means of escape should allow escape from a building in more than one direction and in a direction opposite to that of the main entrance to the building.

30.24 How do I know if the means of escape in my building or site are adequate?

In a modern building it is likely that its construction conforms to the requirements of the Building Regulations and its means of escape may therefore be assumed to be satisfactory. Similarly, if in the course of a recent inspection by the fire service, the means of escape were found to be satisfactory, then nothing further need be done.

If neither of the above apply to you and you are unsure as to whether you have a suitable number of means of escape, there are a few guidelines that will help you to decide on the adequacy of the means of escape:

- In the event of a fire, everyone present must be able to reach a place of safety as quickly as possible.
- Routes and exits must lead as directly as possible to a place of safety.
- The number and distribution and dimensions of emergency routes and exits must be adequate with regard to:
  - The use of the premises
  - The maximum number of people who may be present
  - The dimensions of the emergency routes and exits
  - The fire hazards that are present.
- Emergency doors must open in the direction of travel.
- Sliding or revolving doors must not be used for exits specifically intended as emergency routes.
- Emergency doors must not be locked or fastened that they can not be easily and immediately opened by anyone who requires to use them in an emergency.

- Emergency routes and exits must be indicated by signs.
- Emergency routes and exits that require illumination must be provided with emergency lighting of adequate intensity in the event that the normal lighting fails.

Although it is not stated in the Fire Safety Order it goes without saying that routes to emergency exits and the doorways themselves must be kept clear at all times.

30.25 Are there any exit routes which are not suitable to be designated as means of escape?

Any route which could be hazardous to use, would be deemed to be unsuitable. In particular, the following are considered to be unsuitable as means of escape:

- Lifts
- Portable ladders
- Long spiral staircases
- Escalators
- Throw out devices or escape ropes, etc.

30.26 Why can I not use a lift as a means of escape?

Lifts are unsuitable as a means of escape because:

(1) The power to the lift may be disconnected because of the fire. In addition, occupants of the lift car could be trapped in the lift shaft in a burning building.
(2) The lift could stop and discharge occupants onto the floor on which the fire is burning. The occupants of the lift will not know which floor has the fire burning and could end up walking into it and then become trapped.

There are some designs of lifts which can be used if there is a fire and these are commonly known as "fire fighter" lifts. They are designed to enable them to run on separate power systems and are installed for the use of fire fighters so that they can get to the scene of the fire in a tall building safely and quickly.

30.27 What is a place of safety?

A place of safety is a place in which a person is no longer at risk from a fire. A place of safety must be in the open air and must be a place away from the building. Means of escape are designed to take people to a place of safety that is to a location away from the building where they will not be at risk from heat, fumes, smoke, collapsing structures, etc.

It is not enough for occupants to egress from the building into an adjoining alleyway, small yard, garden or inner courtyard, from which they will not be able to escape and would still be at risk from the fire.

30.28 Is the assembly point also the place of safety?

Not always – the means of escape are designed to lead people to a place of safety i.e. an area in the open air which is away from the building.

However an assembly point is a location at which people are requested to meet so that a roll call can be made to ensure that all occupants of the building have been accounted for.

30.29 What is taken into consideration when designing means of escape?

The first key principle of any means of escape is that people should be able to turn their back to a fire and leave the building to a place of safety.

It is recommended that every building has two means of escape, but this is not always possible and some smaller buildings may only have one means of escape.

Dead ends should be avoided as people may get to the end of the corridor and have no other way to go. This could mean that they are either trapped or have to turn back towards the fire to try and exit the building.

The means of escape should be designed to be wide enough to take the occupants of the area without crushing and overcrowding.

Exits from the building should be separate and independent and not close together.

Exit routes from upper floors will not always lead to an outside place of safety, in this instance an exit route must lead to a place of relative safety.

30.30 What is a place of relative safety?

Where there is a considerable distance between an exit route and the final place of safety, it is necessary to provide an intermediate place of safety, this is known as a place of relative safety. This can be in a location where people can rest for a few seconds before they finally exit the premises.

Multi-storey buildings often have long distances to travel from upper floors to the ground floor exit routes and places of relative safety are often provided on protected staircases, which are safe to remain in whilst waiting to exit the premises.

30.31 Do means of escape corridors have to be sub-divided so as to prevent the spread of fire?

Corridors which exceed 30 metres in length in shops and 45 metres in length in offices and factories, should be sub-divided by fire doors to prevent the free travel of smoke, heat and toxic fumes.

The sub-division of corridors should be designed so that no individual length of corridor is common to more than one storey exit.

30.32 What is meant by the term "the three stages of escape"?

Means of escape can be divided into three stages as follows:

Stage 1   Travel within a room.
Stage 2   Horizontal travel to an exit from an upper floor or to a final exit on the ground floor.
Stage 3   Vertical travel within a stairway to a final exit.

These three stages of escape are a useful tool when considering the implications of means of escape. Means of escape can become complicated, but breaking them down into manageable parts makes is easier to ensure that the means of escape are suitable for your premises.

When planning or checking your means of escape, you need to consider:

Stage 1   How do I get out of this room and where will I end up, once I leave the room.
Stage 2   How will I get out of this floor or main area and where will I end up.
Stage 3   Where are the staircases, where do the staircases lead to and where will I end up when I use them.

30.33 What is meant by the terms protected staircase and protected lobby?

A staircase is classed as being protected if it is enclosed by fire resisting constructed materials. Fire-resisting materials are any material which will withstand the effects of a fire for a minimum of 30 minutes. Some materials offer greater fire resistance than 30 minutes and these are used in certain locations such as public buildings, etc.

A protected staircase must discharge to a final exit or lead onto a protected route which leads to the final exit and subsequently to a place of safety.

A protected lobby is a fire-resisting enclosure providing access to a protected stairway via two sets of fire-resisting self-closing doors into which no rooms open, other than toilets or lifts.

30.34 What is meant by a protected route?

A route which is protected is the one which has a degree of protection from fire, this is provided by the walls, doors, partitions, ceilings and floors which separate the route from the remainder of the building.

30.35 What do I do if I only have one escape route?

If your premise has a small floor area and a restricted number of floors a single escape route may well be satisfactory if all areas are assessed as being of normal or low fire risk.

30.35.1 Items which should not be located on means of escape

Any items which restrict the width of the escape route or increase the hazard of the area should be allowed on escape routes. Therefore corridors, stairways, landings and lobbies should be kept clear of:

- Portable heaters of all types
- Fixed lights or heaters incorporating flames or radiant bars
- Gas cylinders
- Boilers, unless installed in suitable fire-resisting housings
- Cooking appliances, including toasters and microwave ovens
- Electrical appliances, including photocopiers and shredders
- Vending and games machines.

In addition furniture should be kept to a minimum and should not reduce the width of the route. Any furniture necessary, such as in a reception area should comply with the Furniture and Furnishings (Fire)(Safety) Regulations 1988.

Artificial foliage, shrubs and decorations should only be placed on escape routes if the supplier satisfies you that they have behaved satisfactorily when subjected to the flame ignition source of BS5852: Fire tests for furniture.

30.36 What signs should I put up on a means of escape?

In most workplaces, which are deemed to be of normal or low fire risk, only the means of escape which are not in common use need to be signed with a running man pictogram sign.

However, in any area where the public have access, all means of escape routes should be signed. Signs should be displayed clearly along the length of the route so that it may be followed easily and without delay.

Means of escape should no longer be signed just as a "Fire Exit" – any signage should be of the pictogram type, although the signs can incorporate wording along with the pictures if necessary.

Examples of signage which would be suitable for a means of escape route are as follows:

30.37 When do I need to apply the principles of travel distances, protected routes, inner rooms, etc. to existing buildings?

If you are considering internal alterations or refurbishment works to your building you must take into consideration the fire safety requirements which relate to the alterations.

Even tasks such as erecting internal partitions can create what is known as an inner room which leads off an access room. Partitions can also obstruct means of escape routes by increasing travel distances.

Before starting any works you should consider the existing means of escape and plan the remedial works in relation to the principles of good fire safety.

30.38 What are fire doors?

Doors which are fire resisting and self-closing are usually referred to as fire doors. They are a vital part of the building's fire defence system as they are designed to hold back fire and smoke.

Fire doors are usually provided to carry out one of the following functions:

(1) To prevent the integrity of a structural fire compartment by holding back fire and smoke.
(2) To protect the means of escape for the occupiers of the building for a sufficient period of time for them to escape from the building.

Fire doors can provide fire protection which ranges from 30 minutes to 4 hours, depending on the design of them.

Fire doors must be close fitting and constructed of a material which will withstand heat and flames for the period specified by the fire resistance of the door e.g. 30 minutes.

Fire doors are normally manufactured with intumescent strips around the door edge. These strips are designed to act as a smoke seal which will expand when they get hot to fill the gaps between the door and the door frame.

30.39 What is a Fire Safety Plan?

The Fire Safety Plan should identify fire risks throughout the site e.g.:

- Combustible materials
- Use of hot flame equipment
- Use of liquid petroleum gas
- Use of combustible substances
- Storage and use of any explosive materials and substances
- Sources of ignition e.g. smoking
- Use of heaters.

Once the potential fire risks are identified i.e. where, when, why and how a fire *could* start on site (or the surrounding area, yards, outbuildings) the Fire Safety Plan should include precautions and procedures to be adopted to *reduce* the risks of fire. These could include:

- Operating a Hot Works Permit system
- Banning smoking on site in all areas other than the approved mess room
- Controlling and authorizing the use of combustible materials and substances
- Providing non-combustible storage boxes for chemicals
- Minimizing the use of liquid petroleum gas and designating external storage areas
- Controlling the siting and use of heaters and drying equipment
- Operating a Permit to Work system for gas and electrical works.

Having identified the potential risks and the ways to minimize them, there will always be some residual risk of fire. The Fire Safety Plan should then contain the Emergency Procedures for dealing with an outbreak of fire, namely:

- Types and location of Fire Notices
- The location, number and type of fire extinguishers provided throughout the site
- The means of raising the alarm
- Identification of fire exit routes from the site and surrounding areas
- Access routes for emergency services

- Procedure for raising the alarm
- Assembly point/muster point.

The Fire Safety Plan should also contain the procedures to be taken on site to protect against arson e.g.:

- Erection of high fencing/hoarding to prevent unauthorized entry
- Fenced or caged storage areas for all materials, particularly those combustible
- Site lighting e.g. infra-red, PIA
- Use of CCTV
- Continuous fire checks of the site, particularly at night if site security is used.

Procedures for the storage and disposal of waste need to be included as waste is one of the highest sources of fire on construction sites.

Materials used for the construction of temporary buildings should be fire protected or non-combustible whenever possible e.g. 30-minute fire protection. The siting of temporary buildings must be considered early in the site planning stage as it is best to site them at least 10 meters away from the building being constructed or renovated.

Having completed the Fire Safety Plan, a sketch plan of the building indicating fire points, assembly point, fire exit routes, emergency services access route to site, etc. should be completed and attached to the Plan. The sketch plan (which could be an Architect's outline existing drawing) should be displayed at all fire points and main fire exit routes and must be included in any site rules/information handed out at induction training.

30.40 What are general fire precautions on a construction site?

The term general fire precaution is used to describe the structural features and procedures needed to achieve the overall aim of fire safety which is to ensure that:

"Everyone reaches safety if there is a fire"

Putting a fire out is secondary to "life safety".

Fire Risk Assessments are about ensuring that you have considered the likelihood of a fire starting, who it would affect, how quickly and how those people would be evacuated to a place of safety.

General fire precautions cover:

- Escape routes and fire exits
- Fire fighting equipment
- Raising the alarm
- Making emergency plans
- Limiting the spread of fire (compartmentation).

General fire precautions will invariably differ from site to site depending on the complexity of the site.

Life safety for all persons on the site must be properly considered, planned, implemented and regularly checked. Escape routes for instance should be permanent, well identified, leads to a place of safety, well sign posted, unobstructed. An ad-hoc scramble down ladders for jumping off floors will not be acceptable.

30.41 What are some of the basic fire safety measures?

The Fire Safety Plan should be developed as an integral part of the Construction Phase Health and Safety Plan or overall work plan if CDM Regulations do not apply.

The very basics are:

- Fire exit routes from the site
- Fire fighting equipment
- Methods to raise the alarm
- Lighting
- Signage
- Fire protection measures to prevent spread
- Safe systems of work
- Prohibiting smoking on site
- Appointment of Fire Wardens
- Regular checks
- Emergency procedures.

30.41.1 Fire exit routes

Where possible, there should always be more than one exit route from a place of work. If the travel distance is more than 45 metre to an exit, there must be two or more exits. This travel distance will vary depending on the risk rating of the site.

If the number of exits cannot be improved then the risk rating of the site for fire must be reduced.

Fire exit routes must be unobstructed, clearly defined, of adequate size and width, not locked.

Doors leading onto fire exit routes should open via a push bar in the direction of travel.

No fire exit route shall lead back into the building or site.

Fire exit routes must lead to a place of safety.

Fire exit routes must in themselves be protected from fire by fire-protected enclosures or doors. Doors must be kept shut.

Exit routes using ladders e.g. on scaffolding, need to be especially assessed as part of the site-specific Risk Assessment.

Fire exit routes must be clearly visible from all parts of the work area. Exit signs which meet the Health and Safety (Safety Signs and Signals) Regulations 1996 must be displayed.

If lighting is poor – use photo-luminescent signs.

All fire signage must display pictograms as a minimum. Text can also be used alongside the pictogram, as can directional arrows.

30.41.2 Fire fighting equipment

Suitable fire extinguishers need to be placed in appropriate locations around the site, and always at fire points near the fire exit routes.

Multi-purpose foam or powder extinguishers are suitable, but so too would be water and carbon dioxide. The Fire Risk Assessment should determine which type is required.

Fire fighting equipment should be visible, properly signed, inspected weekly and ready to use if needed. Operatives should not need to climb over materials, move plant, etc. to use the extinguishers.

Either a designated number of operatives in each work area or all operatives should be trained in how to use the fire fighting equipment.

Regular re-assessment of the working area is needed to ensure that the location of the fire extinguisher points are appropriate.

30.41.3 Methods of raising the alarm

A fully integrated alarm system would be beneficial on all sites, activated by break glass points and linked to an alarm control panel in the site office.

However, this is not always possible and alternatives are permissible e.g.:

- Hand bells
- Klaxons
- Sirens
- Hooters.

The alarm in use on the site should be clearly identified and all operatives *must* receive training in fire alarm procedures.

Fire alarm points must be clearly visible, easily accessible, etc.

If alarms cannot be heard in all areas of the site there must be a procedure for Fire Wardens to warn other floor Fire Wardens, etc.

On small sites, a simple shouting of Fire! Fire! May be all that is needed.

If the site is multi-occupied without employers e.g. major department store refurbishment then the construction site alarm system must integrate with that of other employers so that total building evacuation is occasioned as necessary.

30.41.4 Lighting

Emergency lighting is not necessarily required on all construction sites but if there is a risk of power failure and no natural daylight to the areas of work, emergency lighting will be essential.

A simple system of torches may suffice.

All emergency exit routes must be adequately lit at all times.

Emergency backup lighting will activate when the main power supply fails.

Regular checks of emergency lighting will be necessary.

30.41.5 Signage

Signage enables people to be guided to safe places – either to emergency exit routes, safe places e.g. refuges or to assembly points.

Signs where possible could be photo-luminescent.

Signs must be visible from all work areas, non-confusing, of large enough size and accurate in the information they portray e.g. must not lead to a dead end as a fire exit route.

30.41.6 Fire protection measures to prevent fire spread

Generally, it is best to try to consider floors and different areas as compartments, with fire protected, closed doors, stopping and fire protection to voids and ducts, etc.

Fire and smoke, including toxic fumes, spread rapidly. Compartmentation constrains it to one area.

30.41.7 Safe systems of work

Any work activity which looks as if it could increase the risk of a fire starting must be controlled by a Permit to Work or Hot Works Permit system.

Hot works should be prevented whenever possible. Controls need to be localized e.g. additional fire extinguishers, regular checks, additional Fire Wardens.

Combustible materials e.g. flammable gases and so on should be removed. Flashover should be considered.

Only trained operatives should carry out hot works or use flammable materials, etc.

30.41.8 Smoking on site

There is no other fire safety procedure acceptable other than to ban it completely.

However, it may be permitted in mess rooms. If so, strict controls must be implemented.

30.41.9 Appointment of Fire Wardens

Each floor or work area should have an appointed Fire Warden i.e. people who are trained to know what to do in the event of a fire and to evacuate their area, raise the alarm, etc.

There should be enough Fire Wardens to cover for absences. Fire Wardens should receive regular training.

30.41.10 Regular checks

Daily and weekly fire safety checks are advisable on all sites. Checks are always necessary after hot works, and usually approximately 1 hour after the end of hot works so that any smouldering materials can be identified.

Records of fire safety checks should be kept for the duration of the project. Remember, you need to demonstrate that you know what you are doing.

30.41.11 Emergency procedures

These must be specific for each site and written down clearly. Emergency procedures must be displayed in prominent positions.

They should include:

- Type of fire alarm
- How to raise the alarm
- How to evacuate the site
- The assembly point
- The names of the Fire Wardens
- Any highly hazardous areas
- Storage of flammable materials
- Procedures for visitors to site
- Name and telephone numbers of local emergency services
- Liaison with emergency services when they arrive on site.

 Top Tips

- Plan fire safety before works start
- Reduce combustible materials
- Reduce ignition sources
- Keep fire exit routes clear
- Display adequate and suitable fire signage
- Train operatives in emergency procedures
- Keep fire extinguishers on site, in suitable locations and of the correct type
- Put in emergency lighting if possible
- Clearly describe the fire alarm raising procedure
- Remember – get everyone out rather than fight the fire.

Fire Safety Checklist

Identifying sources of ignition

- Smokers materials
- Naked Flames
- Electrical, gas or oil fired heaters, fixed or portable
- Hot processes such as welding, grinding work or cooking
- Engines or boilers
- Machinery
- Faulty or misused electrical equipment
- Lighting equipment such as halogen lamps
- Hot surfaces and obstruction of equipment ventilation
- Friction from drive belts, etc.
- Static electricity
- Metal impact such as metal tools striking each other
- Arson.

Identifying sources of fuel

- Flammable liquid-based products such as paints varnish thinners and adhesives.
- Flammable liquids and solvents such as petrol, white spirit, methylated spirit and paraffin.
- Flammable chemicals.
- Wood.
- Paper and card.
- Plastics, rubber and foam such as polystyrene and polyurethane, e.g. the foam used in upholstered furniture.
- Flammable gases such as liquefied petroleum gas (LPG) and acetylene.
- Furniture, including fixtures and fittings.
- Textiles.
- Loose packaging material.
- Waste materials, in particular finely divided materials such as wood shavings, off-cuts, dust, paper and textiles.
- Hardboard, chipboard, blockboard walls or ceilings.
- Synthetic ceiling or wall coverings, such as polystyrene tiles.

Identifying sources of oxygen

- Natural airflow through doors, windows and other openings.
- Mechanical air conditioning systems and air handling systems.
- Some chemicals (oxidizing materials), that can provide a fire with additional oxygen and so, assist it to burn. These chemicals should

Some people often experience temporary deafness after leaving a noisy place e.g. after leaving a noisy nightclub or bar. Although this type of hearing loss usually results in the hearing recovering within a few hours, this should not be ignored. Temporary hearing loss is a sign that if the person continues to be exposed to the noise their hearing could be permanently damaged.

Permanent hearing damage can be caused immediately by sudden, extremely loud, explosive noises, e.g. from guns or cartridge operated machines, but tends to be a gradual process.

It may only be when damage caused by noise over the years combines with hearing loss due to ageing that people realize how deaf they have become.

Hearing loss is not the only problem. People may develop tinnitus (ringing, whistling, buzzing or humming in the ears). Anyone of any age can suffer from hearing damage.

31.13 What is a low noise purchasing policy?

This basically means that when new equipment is bought or hired the employer will ensure that the quietest equipment or machinery is bought or hired. The cost of introducing noise reduction measures is often reduced quite significantly if quiet equipment can be introduced into the workplace.

31.14 What should be included in the low noise purchasing policy

The following are a few tips which may be helpful:

- Consider at an early stage how new or replacement machinery could reduce noise levels in the workplace – set a target to reduce the noise levels if possible.
- Ensure a realistic noise output level is specified for all new machinery, and check that tenderers and suppliers are aware of their legal duties.
- Ask suppliers about the likely noise levels under particular conditions in which the machinery will be operated, as well as under standard test conditions. (Noise output data will only ever be a guide as many factors affect the noise levels experienced by employees).
- Only buy or hire from suppliers who demonstrate a low-noise design, with noise control as a standard part of the machine.
- Keep a record of the decision process which was followed during the buying or hiring of new machinery, to help show that legal duties to reduce workplace noise have been met.

31.15 What are manufacturers and suppliers of machinery required to do?

Under the Health and Safety at Work etc Act 1974 and the Supply of Machinery (Safety) Regulations 1992 (as amended) a supplier of machinery must do the following:

- Provide equipment that is safe and without risk to health, with the necessary information to ensure it will be used correctly.
- Design and construct machinery so that the noise produced is as low as possible.
- Provide information about the noise the machine produces under actual working conditions.

New machinery must be provided with:

- A "Declaration of Conformity" to show that it meets essential health and safety requirements.
- A "CE" mark.
- Instructions for safe installation, use and maintenance.
- Information on the risks from noise at workstations, including:
  - A-weighted sound pressure level, where this exceeds 70 dB
  - maximum C-weighted instantaneous sound pressure level, where this exceeds 130 dB
  - sound power (a measure of the total sound energy) emitted by the machinery, where the A-weighted sound pressure level exceeds 85 dB.
- A description of the operating conditions under which the machinery has been tested.

31.16 When should hearing protection be used?

If noise cannot be controlled by other methods, such as new machinery, acoustic screening, provision of anti-vibration mounts to machinery or change in working patterns for example, then extra protection for employees will be needed. In addition it may be needed as a short-term measure while other methods of control are being implemented.

Hearing protection should not be used as an alternative to controlling the noise by technical or organizational methods. It should only be used where there is no alternative way of protecting employees from noise exposure.

31.17 What are the general requirements for hearing protection?

- If employees ask for hearing protection, it should be provided for them.
- It must be made available for employees to use when the lower action level of 80 dB(A) is exceeded.
- It must be used by employees when the upper action level of 85 dB(A) is exceeded.
- Employers must provide training and information in the correct use of the hearing protection.
- Employers must ensure that any hearing protection which is provided is properly used and maintained.

31.18 What should maintenance of hearing protection involve?

Any hearing protection which is provided for employees should be checked and maintained on a regular basis to ensure it remains in good working order. As a minimum the employer should check that:

- It is in good condition and clean
- The seals on ear muffs are not damaged
- The tension on headbands of ear defenders is still good and fits well when worn
- Employees have not made any modifications to the hearing protection which may mean that it does not provide the correct level of protection
- Earplugs are still soft, pliable and clean.

31.19 What else can be done?

It is best practice to include the wearing, maintenance and care of hearing protection in the Company Safety Policy.

Managers and supervisors should be encouraged to set a good example by ensuring that they wear hearing protection in areas where it is required.

The hearing protection used must give enough protection, as a guide it should give enough protection to get below 85 dB.

Employers should ensure that the hearing protection is suitable for the working environment and consideration should be given to how hygienic they are and how comfortable they are.

If other personal protective equipment is worn by the employee, such as dust masks or hard hats the employer should consider how the hearing protection will fit in with the other protection. For example the wearing of ear defenders may be difficult if the employee also has to wear a hard hat, in this case a hard hat incorporating a set of ear defenders would be required.

It is important to ensure that the hearing protection provided is not:

- Designed to cut out too much noise, as this can cause isolation and lead to employees not wanting to wear them
- Compulsory to wear, unless the law requires it.

31.20 What information, instruction and training do employees need?

Employees need to understand the risk that they may be exposed to from noise in the workplace. If employees are exposed to the lower action level of 80 dB, as a minimum employers should tell them:

- The estimated noise exposure and the risk to hearing.
- What is being done to control the risks and exposure, this may include the use of acoustic screening, increased maintenance of noisy pieces of equipment, change in work patterns, or provision of hearing protection.
- Where they can get hearing protection from.
- Who in the company will be responsible for the provision and maintenance of the hearing protection and who they should report defects to.
- The correct way to use the hearing protection, how to look after it, how it should be stored and the area where it needs to be used.
- Details of any health surveillance programme which may be in place.

31.21 What is health surveillance for noise?

Health surveillance for hearing usually means:

- Regular hearing checks in controlled conditions by a trained professional.
- Telling employees about the results of their hearing checks.
- Keeping health records.
- Ensuring employees are examined by a doctor where hearing damage is identified.

Ideally health surveillance should be started before employees are exposed to noise, this helps to give a baseline and can be used to determine any changes in noise exposure and the effects those changes may be having.

31.22 What is the purpose of health surveillance?

The health surveillance is designed to provide the employer with an early warning system as to when employees might be suffering from early signs of hearing damage. It gives the employer an opportunity to do something

32

Training in Health and Safety

32.1 What is an employer responsible for in respect of training employees whilst they are at work?

The Health and Safety at Work Etc Act 1974, Section 2, sets out the duties of employers as:

> *"the provision of such information, instruction, training and supervision as is necessary to ensure, so far as is reasonably practicable, the health and safety at work of his employees".*

Training should be considered as a risk control measure for hazards identified within the workplace.

The employer can provide his employees with any combination of instruction, information, training and supervision as is appropriate.

The Management of Health and Safety at Work Regulations 1999, Regulation 13, states that an employer must:

- Provide training upon recruitment and induction
- Provide training whenever an employee is exposed to new or altered risks in respect of people, machinery, processes, materials, etc.
- Provide continuous, repeated training so that employees are given information on current best practice
- Provide training in methods which are flexible and adaptable and which meets the needs of special groups of workers e.g. those with disabilities, literacy problems, language problems
- Provide training in working hours as a business necessity and without charge to employees.

32.2 What are the "five steps" to information, instruction and training?

The HSE publishes many guidance documents on health and safety matters and in its useful "five steps" series it covers training as:

Step 1: Determine who needs training
Step 2: Decide what training is needed and what the objectives are
Step 3: Decide how the training should be given, carried out
Step 4: Decide when training should be carried out
Step 5: Check that the training has worked.

Generally, all employers need to have a policy on health and safety training i.e.:

- What you are going to do
- When you are going to do it
- What subjects will be covered
- Who will do it
- How often
- What assessment tests will be made.

Increasingly, training records are vitally important to prove that as an employer, you have discharged your duties. If an accident occurs the first documents the investigating officer will want to see will probably be:

- Risk assessments
- Training records.

Employees who have been able to demonstrate that they have not been given adequate training are more likely to be successful in civil claims than those who have been trained.

Case Study

A contractor's employee had his hand crushed between the lid of a container and its frame as the lid crashed down whilst he was using it. The HSE Inspector was planning to prosecute for the accident i.e. an unsafe system of work because the container lid had not been secured in position. But the contractor was able to demonstrate that the employee had been given training and did know how to secure the lid into the upright position. He chose not to do it through carelessness. The HSE Inspector did not prosecute because the employer had suitable and sufficient risk assessments and had training records that proved that he had implemented a robust tool box talk training programme for all of his workers.

32.3 What are the legal requirements for instruction and training under the various health and safety regulations?

32.3.1 General health and safety

- **Management of Health and Safety at Work Regulations 1999**
  Health and Safety training:
  - On recruitment
  - On being exposed to new or increased risks
  - Repeated as appropriate.
- **Health and Safety (First Aid) Regulations 1981**
  First aiders provided under the regulations must have received training approved by the HSE.
- **Health and Safety (Safety Signs and Signals) Regulations 1996**
  Each employee must be given instruction and training on:
  - The meaning of safety signs and
  - Measures to be taken in connection with safety signs.
- **Health and Safety (Consultation with Employees) Regulations 1996**
  Training for employee representatives in their functions as representatives (as far as is reasonable). You are required to meet the costs of this training, including travel and subsistence and giving time off with pay for training.
- **Safety Representatives and Safety Committees Regulations 1977**
  Sufficient time off with pay for safety representatives to receive adequate training in their functions as safety representative.
- **Control of Substances Hazardous to Health Regulations 2002**
  Instruction and training in:
  - Risks created by exposure to substances hazardous to health (e.g. high hazard biological agents) and precautions
  - Results of any required exposure monitoring
  - Collective results of any required health surveillance.
- **Health and Safety (Display Screen Equipment) Regulations 1992**
  Adequate health and safety training in the use of any workstation to be used.
- **Control of Noise at Work Regulations 2005**
  Instruction and training for employees likely to be exposed to daily personal noise levels at 85 ldB(A) or above:
  - Noise exposure: level, risk of damage to hearing and action employees can take to minimize that risk
  - Personal ear protectors (to be provided by employers): how to get them, where and when they should be worn, how to look after them and how to report defective ear protectors/noise control equipment
  - When to seek medical advice on loss of hearing
  - Employees' duties under the regulations.

- **Control of Asbestos Regulations 2006**
  Instruction and training about risks and precautions for:
  - Employees liable to be exposed to asbestos
  - Employees who carry out any work connected with your duties under these regulations.
- **Control of Lead at Work Regulations 2002**
  Instruction and training about risks and precautions for:
  - Employees liable to be exposed to lead
  - Employees who carry out any work connected with your duties under these regulations.
- **Ionising Radiations Regulations 1999**
  Instruction and training to enable employees working with ionizing radiation to meet the requirements of the regulations e.g. in radiation protection for particular groups of employees classified in the regulations.
- **Provision and Use of Work Equipment Regulations 1998**
  Employees who use work equipment (including hand tools) and those who manage or supervise the use of work equipment need health and safety training in:
  - Methods which must be used
  - Any risks from use and precautions.
- **Personal Protective Equipment at Work Regulations 1992**
  Employees who must be provided with PPE need instruction and training in:
  - Risk(s) the PPE will avoid or limit
  - The PPE's purpose and the way it must be used
  - How to keep the PPE in working order and good repair.

32.4 What instruction and training needs to be given on fire safety?

Fire training to the extent that employees should know that action to take when fire alarms sound, should be given to all employees and should be included in the induction training. Knowledge of particular emergency plans and how to tackle fires with equipment available may be given in specific training at the workplace. At whatever point training is given, the following key points should be covered:

- Evacuation plan for the building in case of fire, including assembly points(s)
- How to use fire fighting appliances available
- How to use other protective equipment, including sprinkler and other protection systems, and the need for fire doors to be unobstructed
- How to raise the alarm and operate the alarm system from call points
- Workplace smoking rules
- Housekeeping practices which could permit a fire to start and spread if not carried out e.g. waste disposal, use of ash bins, handling of flammable liquids and so on.

Fire training should be accompanied by practices, including regular fire drills and evacuation procedures. No exceptions should be permitted at these.

32.5 What specifically must employers provide in the way of information on health and safety issues to employees?

- **Management of Health and Safety at Work Regulations 1999**
  Information on:
  - Risks to health and safety
  - Preventative and protective measures
  - Emergency procedures including evacuation
  - Specific health and safety risks for temporary employees
  - Requirements for any health surveillance
  - Competent persons
  - Risks created by other employers.
- **Control of Substances Hazardous to Health Regulations 2002**
  Information on:
  - Risks to health created by exposure to substances hazardous to health (including e.g. high hazard biological agents)
  - Precautions
  - Results of any required exposure monitoring
  - Collective results of any required health surveillance
  - Safety data sheets.
- **Chemicals (Hazards Information and Packaging) Regulations 2002**
  Safety data sheets or the information they contain to be made available to employees (or to their appointed representatives).
- **Manual Handling Operations Regulations 1992**
  Information on:
  - The weight of loads for employees undertaking manual handling and
  - The heaviest side of any load whose centre of gravity is not positioned centrally.
- **Health and Safety (Display Screen Equipment) Regulations 1992**
  Health and safety information about display screen work for both operators and users (the regulations define who is an operator and who is a user).
- **Health and Safety (First Aid) Regulations 1981**
  First aid arrangements: including facilities, responsible personnel and where first aid equipment is kept.
- **Health and Safety (Safety Signs and Signals) Regulations 1996**
  Each employee must be given clear and relevant information on the measures to be taken in connection with safety signs.

- **Health and Safety Information for Employees Regulations 1989**
  Information about employees' health, safety and welfare in the form of:
  - An approved poster to be displayed where it can be easily read as soon as is reasonably practicable after any employees are taken on
  - An approved leaflet to be given to employees as soon as practicable after they start.
- **Health and Safety (Consultation with Employees) Regulations 1996**
  Necessary information to enable your employees to fully take part in consultation and to understand:
  - What the likely risks and hazards arising from their work, or changes to their work, might be
  - The measures in place, or to be introduced, to eliminate or reduce them
  - What employees ought to do when encountering risks and hazards.
- **Safety Representatives and Safety Committees Regulations 1977**
  Necessary information to assist the work of safety representatives nominated in writing by a recognized trade union.
- **Ionising Radiations Regulations 1999**
  Information:
  - To enable employees working with ionizing radiations to meet the requirements of the regulations
  - On health hazards for particular employees classified in the regulations, the precautions to be taken and the importance of complying with medical and technical requirements
  - For female employees on the possible hazard to the unborn child and the importance of telling the employer as soon as they find out they are pregnant.
- **Control of Pesticides Regulations 1986**
  Information on risks to health from exposure to pesticides and precautions.
- **Provision and Use of Work Equipment Regulations 1998**
  Information on:
  - Conditions and methods of use of work equipment (including hand tools) and
  - Foreseeable abnormal situations; what to do and lessons learned from previous experience.
- **Personal Protective Equipment at Work Regulations 1992**
  Information on:
  - Risk(s) that the PPE will avoid or limit, and
  - The PPE's purpose and the way it must be used
  - What your employee needs to do to keep the PPE in working order and good repair
- **Control of Asbestos Regulations 2006**
  Information about risks and precautions for:
  - Employees liable to be exposed to asbestos
  - Employees who carry out any work connected with your duties under these regulations.

- **Control of Lead at Work Regulations 2002**
  Information about risks and precautions for:
  – Employees liable to be exposed to lead
  – Employees who carry out any work connected with your duties under these regulations.
- **Control of Noise at Work Regulations 2005**
  Information on:
  – Risk of damage to hearing
  – What steps are to be taken to minimize risk
  – Steps the employee must take to obtain personal ear protection
  – Employees obligations.

32.6 Are tool box talks a suitable way of training operatives on a construction site?

Yes – tool box talks can be very effective in getting key messages across to a range of contractors.

Tool box talks are short duration – approximately 5–10 minutes – presentations which focus on just one topic and update/remind or inform operatives about key health and safety messages which should be practiced on a construction site.

Good training is provided in a medium which is understood by the recipients – i.e. verbal instructions, posters, quiz games, videos and so on may all be a much better presentation method than a long lecture.

Training should be practical and commensurate with the needs of the job e.g. there is no need to train an electrician about the hazards of steel erection if he will never undertake this role, but an electrician may need to know that there might be a hazard to his safety whilst steel girders are moved around or delivered to site.

Tool box talks can be supported by pictorial aids and this will help overcome any language barriers.

The Principal Contractor or contractor should keep a record of all those who attend tool box talk sessions, the topic and date of training.

Records are important as they demonstrate the steps taken by the employer, or other person as appropriate, to comply with their legal duties of "information, instruction and training".

APPENDICES

Appendix 5B

STATEMENT OF APPOINTMENT
CDM CO-ORDINATOR

We, [CLIENT], do hereby appoint [CDM CO-ORDINATOR'S NAME], of [CDM CO-ORDINATOR'S ADDRESS], to the position of CDM Co-ordinator as required by the Construction (Design and Management) Regulations 2007.

The CDM Co-ordinator will supply the services as listed in the Specification of Duties attached to this statement of appointment.

[CDM CO-ORDINATOR] shall assume their automatic appointment to all notifiable projects (as defined in CDM 2007) and will not receive individual letters of appointment. This appointment will run until terminated by either party.

The position of CDM Co-ordinator is assumed until the final handover of a project from the Principal Contractor to [CLIENT].

In making this appointment, [CLIENT] confirm that they are fully aware of the duties imposed on them by CDM 2007 and understand the role that the CDM Co-ordinator must perform.

Signed: _____ Signed: _____

      For [CLIENT]   For [CDM CO-ORDINATOR]

Date: _____ Date: _____

Appendix 5C

CONSTRUCTION (DESIGN AND MANAGEMENT) REGULATIONS 2007

CLIENT DUTIES AND F10 AUTHORIZATION

We, [Client], confirm that we are aware of and understand, the following duties placed on Clients in the CDM Regulations 2007:

- Check competence and resources of all appointees, including designers, contractors, CDM Co-ordinators and other works involved in construction projects.
- Ensure co-operation and co-ordination of all persons involved in the project so far as it relates to our duties as Client.
- Ensure that there are suitable management arrangements in place for the project, in particular in relation to:
  - Construction work to be undertaken without risk to health and safety of any person
  - Compliance with Schedule 2 of the Regulations (welfare facilities)
- Allowing sufficient time and resources for planning and preparation of all stages of the project
- Providing pre-construction information
- Appointing a CDM Co-ordinator for notifiable projects
- Appointing a Principal Contractor for notifiable projects
- Ensuring that construction work does not start on notifiable projects until a construction phase health and safety plan is in place
- Providing information for the Health and Safety File to the CDM Co-ordinator
- Retaining and providing access to the Health and Safety File
- Ensuring that procedures are in place to demonstrate that structures designed to be used as a workplace comply with the Workplace (Health, Safety and Welfare) Regulations 1992
- Ensuring that arrangements made for managing the project are maintained and reviewed throughout the project.

In addition to the above, we acknowledge that it is the duty of the Client to sign the F10 declaration sent to the HSE for all notifiable projects or to authorise another to sign it on our behalf provided they are confident that we understand our duties.

[Client] hereby authorise [name of CDM Co-ordinator] to sign all F10's which relate to construction works, new build works, refurbishment and repair works to our premises which operate under the names of [Client name].

Signed: _____
                Company and Position

Date: _____

Appendix 5D

[CLIENT NAME]
CDM PERMIT TO PROCEED

Name of Project:	
Address of Project:	
Name of Designers:	
Name of Principal Contractor:	
Name of CDM Co-ordinator:	

The following information is confirmed as being available, and where necessary, checked and approved by the appropriate person:

		YES	NO	N/A
1.	Asbestos survey, asbestos removal and issue of clear air certificates			
	Signed:			
2.	Party wall agreements			
	Signed:			
3.	All necessary planning permissions and Building Control applications			
	Signed:			
4.	Fire Safety Risk Assessments/Plans for existing buildings			
	Signed:			
5.	DDA Compliance – new building			
	Signed:			

6. Welfare facilities as specified in the pre-construction information:
No. of operatives expected on site:
No. of WC's:
No. of urinals:
No. of wash hand basins:
Canteen facilities:

		YES	NO	N/A
7.	F10 completed and sent to HSE			
8.	Management arrangements for health and safety on site State name of site supervisor/agent:			
9.	Satisfactory completion of the Construction Phase Health & Safety Plan			
10.	Successful design co-ordination and agreement by all parties that best practice has been achieved in respect of health and safety issues relevant to the designs of all elements of the project e.g. construction, M&E, services, end users, maintenance, direct suppliers			

11. Confirmation by designers that their designs meet the requirements of the Workplace (Health, Safety and Welfare) Regulations 1992

Signed:

(Designer)

12. Please comment on any aspect above not completed or confirmed and state whether such omissions should delay the start on site date:

13. Confirmation that adequate time and resources have been allowed for planning and preparation before start on site date i.e. indicate time allowed

_____ Weeks

14. Subject to agreement by all parties, this project is signed off to commence on site on:

_____

Signed: _____ CDM Co-ordinator
_____ Principal Contractor
_____ Principal/Lead Designer

Appendix 5E

[CLIENT NAME]
NON-NOTIFIABLE PROJECTS PERMIT TO PROCEED

CONSTRUCTION (DESIGN AND MANAGEMENT) REGULATIONS 2007

Name of project: _____

Description of works: _____

Date works due to commence: _____

Duration: _____

(NB: must be 30 days or less, or involve 500 person days or less)

CDM Requirements	Yes No N/A	Comments
Approved competent contractors appointed.		
Appropriate pre-construction information about the project, procedures etc issued to contractors, e.g. asbestos location, opening hours, use of staff accommodation, residential hazards, etc.		
Contractors aware of safety procedures e.g. permit to work schemes, evacuation procedures, etc.		
Contractor has produced a written health and safety construction plan commensurate with the risks on the project and has or will communicate to all other contractors.		
Adequate management provided on the site Named site agent:		
Adequate arrangements made for welfare facilities i.e. WC's, wash hand basins, eating of meals, drinking water. Please describe:		
Appropriate co-ordination between designers has taken place and all are aware of and agree with works to be undertaken.		
Design Risk Assessments have been completed and provided to those who will need them.		

Preparation and Planning

The project has been properly planned and resourced and the following timescales have been used:-

Design, planning and preparation: _____ days/weeks

Pre-start preparation i.e. time between
appointment of contractor and start on site _____ days/weeks

All agreed works can start on site: YES/NO/NA

Date: _____

Signed: _____ On behalf of Client

_____ On behalf of Contractor

_____ On behalf Designer

(9)   Describe the general condition of the building e.g. Good (as will probably be for most), run down, dilapidated, etc.

(10)   Is it below ground level if so has it been tanked? Is it on a flood plain or near a river?

(11)   Is there any equipment/plant that needs to be removed during the project?

(12)   If "no" state whether a current asbestos register is available.

(13)   Detail areas where fragile materials may or are located, such as roofs, glazed areas, shopfronts and so on.

(14)   Evidence of pests including droppings, feathers, sight of actual rodents/birds/insects, etc.

(15)   Syringes, other equipment associated with drug abuse.

(16)   Where will skips be located, how operated (weight and load, permanent skip, etc.).

(17)   Is there a likelihood that deliveries will need to be co-ordinated with neighbouring premises. Could deliveries to neighbouring premises (either by nature or size) interfere with operations.

(18)   What is the nature of other works (if any)? Who is the principal contractor?

(19)   Applicable if floor/ground penetration forms part of scope of works.

(20)   Were photographs taken on site?

(21)   E.g. Fan for kitchen extract, etc.

(22)   Confined spaces, temporary supports, weakened structure, areas where a risk of falling over 2 metre exists, etc.

(23)   Steel supports, will propping be required?

(24)   Does the proposed works require lifts/hoists for deliveries or food service. Are there existing lifts/hoists within the area?

(25)   Location of meters/main stop valve, etc.

(26)   Sealed off and not used?

The Construction Phase Health and Safety Plan

When a Principal Contractor is appointed to a project, a site survey will be necessary to establish additional information regarding the building and site in order for the construction phase health and safety plan to be developed.

Key information on hazards and risks regarding the project will be contained in the pre-construction health and safety information pack. The Principal Contractor must develop the pre-construction health and safety information into the construction phase health and safety plan.

A site survey should be used to gather information on:

• Use of the building – will the Client still have employees, customers, etc. within the building?

• What safety rules, risk assessments, etc. are available for the site or activity?

• What are the emergency procedures?

• What restrictions are there to site access?

- What will delivery routes be?
- What residual hazards are on the site/within the building?
- What site security procedures will need to be followed?
- Location of the nearest hospital, etc.
- Location of underground and above ground services.
- Use of adjacent land/buildings.
- Location of site welfare facilities.
- Access for construction workers.
- Storage areas.
- Manual handling hazards e.g. Need to handball materials.
- Location for any cranes, access equipment.
- Risk of unauthorised persons to site.
- Hazardous substances on the site.
- Presence of asbestos.
- Large delivery items and access routes.
- Type of heating required to the site.
- Type and location of lighting.

Much of the above information should be contained in the pre-construction health and safety pack, but if it is not, it does not mean that the hazard, etc. does not apply to the site. The onus is on the principal contractor to assess site safety hazards himself and plan to avoid/control them.

Appendix 10B

PERRY SCOTT NASH
associates limited

Perry Scott Nash House, 2 Arlington Court, Whittle Way, Arlington Business Park, Stevenage, SG1 2FS
t 01438 745 771 f 01438 745 772 e info@perryscottnash.co.uk w www.perryscottnash.co.uk

PRE-CONSTRUCTION INFORMATION PACK
for the purposes of the
CONSTRUCTION (DESIGN AND
MANAGEMENT) REGULATIONS 2007

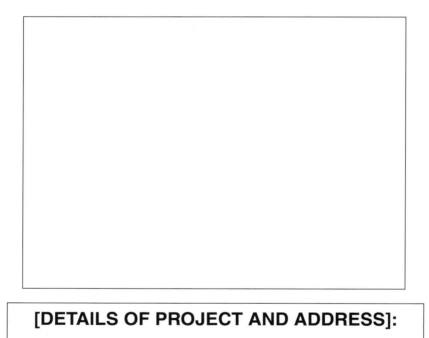

[DETAILS OF PROJECT AND ADDRESS]:

Contents

Part 1 – Site-specific issues

Part 2 – Employers requirements : Health and safety

Part 3 – Construction Phase Health & Safety Plan

Part 1 Site-specific issues

1. Introduction

The Construction (Design and Management) Regulations 2007 require that a health and safety plan is prepared and maintained until the end of the construction phase.

The purpose of this document is to provide the appointed Principal Contractor with the necessary information to allow the project to be undertaken safely. The CDM Co-ordinator has prepared this site-specific document to comply with Regulation 10 and 15 of the CDM Regulations and it should be used by the Principal Contractor to develop the required Construction Phase Health and Safety Plan under Regulation 23.

The successful tenderer will be appointed as the Principal Contractor for this project.

This document consists of three parts:

Part 1 The site-specific safety aspects of this project. This contains specific information and the identification of hazards, with respect to the existing premises and proposed works.

Part 2 The Client's safety requirements for the management of the project. This details the Client's requirements with respect to safety management applicable to all development works for [CLIENT].

Part 3 The Construction Phase Plan – preferred template.

All tendering contractors must allow sufficient resources to comply with the requirements of both sections.

Prepared by: CDM Co-ordinator

Signature: ....

Date: ....

2. Brief description of the project

2.1   Description of works to be carried out

2.2   Description of future use of the building

3. Project directory

Client:

CDM Co-ordinator:

Designers

Architect/Designer:

Quantity Surveyor:

M&E Consultant:

Structural Engineer:

Contractors

Principal Contractor:

Specialists Contractors:

4. Local information directory

**Local Hospital
(with A&E Department):**   Tel:

HSE Office: Tel:

Local Authority: Tel:

RIDDOR Incident Contact Centre: Tel: 0845 300 9923
Fax: 0845 300 9924

Service / Utility Companies: Water:

Gas:

Electricity:

5. Project programme

5.1   F10 notification to HSE:

5.2   Tendering programme: Project out to tender:
Tender return date:
Tender reviews:

5.3   Appointment of Principal Contractor:

5.4   Commencement of Preparation and Planning stage:

5.5   Minimum period for preparation and planning: 4 weeks

5.6   Start of Construction Phase:

5.7   Duration of Construction Phase: weeks

5.8   Contractual completion of project:

5.9   Builders completion:

5.10 Building/project handover:

5.11 Merchandising and on-site training:

5.12 Opening date:

N.B.: Prior to start on site, [CLIENT] will allow a period of ******* for planning and preparation, including site set up.

Within the ******* period the Principal Contractor is to ensure that he has procured sufficient welfare facilities for the site to satisfy the appropriate standards (Schedule 2, CDM 2007).

6. Roles and responsibilities of CDM duty holders

6.1   Client

The Client is [CLIENT NAME] and any authorised representative acting on behalf of the company.

The Client is responsible for complying with their statutory duties under the Construction (Design and Management) Regulations 2007, namely to:-

- Appoint competent persons to all duty posts.
- Ensure adequate resources are available.
- Ensure health and safety management procedures are implemented on projects.
- Ensure adequate welfare facilities are available for projects on site.
- Appoint a CDM Co-ordinator for notifiable projects.
- Appoint a Principal Contractor for notifiable projects.
- Provide appropriate information to those who need it for all projects.

6.2   CDM Co-ordinator

The CDM Co-ordinator is [CDM CO-ORDINATOR NAME]. The Co-ordinator is responsible for:

- Demonstrating that they are competent to carry out the role and that they have the resources for the commission.
- Advising and assisting the Client on arrangements for managing the project.
- Advising and assisting the Client on appointments.

- Advising and assisting the Client on determining if the project is notifiable, and if so, notify the HSE.
- Advising and assisting the Client on the competence, capability and resources of others.
- Providing information, co-ordination and planning, and preparations for project construction work.
- Working with designers on risk reduction and health and safety management.
- Advising and assisting clients on the suitability of the start on-site date.
- Dealing with design work during the construction phase.
- Delivering a suitable Health and Safety File to the Client at the end of the project.

6.3 Designers

Designers on this project will be:
Architects
Quantity Surveyors
Structural Engineers
M&E Consultants

The Principal Contractor will also be a designer if designs are progressed or altered on site.

Designers are responsible for ensuring that they consider all aspects of hazard and risk in their designs and that they provide appropriate information to those who need it on any residual risks.

Designers are referred to their specific duties under CDM 2007, particularly in respect of the duty regarding prevention as specified in Regulation 7.

6.4 Principal Contractor

The Principal Contractor shall have ultimate responsibility for the safe operation of the construction works and shall ensure that they fulfil their duties as set out in Regulations 22, 23 and 24 CDM 2007.

The Principal Contractor shall ensure that they collect all relevant information regarding the works and forward it to the CDM Co-ordinator.

The Principal Contractor shall also undertake the duties of a Designer where they design, specify, alter a design or otherwise act as if designing the structure, etc.

6.5 Contractors

Contractors are all companies, organisations and persons who carry out any works on a [CLIENT NAME] site, including all Client direct appointments.

All contractors are responsible for complying with their duties as set out in Regulation 19 CDM 2007.

7. Existing environment and site

7.1 Existing Premises

7.2 External Areas

7.3 Neighbouring Public Areas

7.4 Local Infrastructure/Utilities

7.5 Access and Deliveries

7.6 Hazardous Areas

7.7 Environment/Site-Specific Safety Requirements

8. Site survey information

8.1 Asbestos

8.2 Contaminated Land

8.3 Hazardous Substances

8.4 Radon Gas/Mining Reports

8.5   Timber and Damp

8.6   Condition/Dilapidation Survey

8.7   Structural Stability Survey

8.8   Pigeon, Rodent, Pest Control Survey

8.9   Utilities Survey

8.10 Acoustic Consultants Report

8.11 Party Wall Surveyor Report

8.12 Archaeological Survey

8.13 Water Contamination Survey

8.14 Other Survey Information

9. Existing information

9.1 The following information is available from the sources indicated below:

Information Type   **Source**

- Existing Drawings
- Health & Safety File
- Proposed Layout and Design Drawings
- Structural Engineer Drawings
- Mechanical and Electrical Drawings
- Utilities Location Drawings
- Asbestos Survey Report
- Asbestos Register
- Hazardous Substances Reports (see previous section)
- Condition/Dilapidation Report
- Planning Consent
- Building Regulations Approval
- Fire Risk Assessment

The recommended sanitary provisions are scaled in the table below:

No. of Men at Work	No. of Water Closets	No. of Urinals	No. of Wash Stations
1–15	1	1	2
16–30	2	1	3
31–45	2	2	4
46–60	3	2	5
61–75	3	3	6
76–90	4	3	7
91–100	4	4	8

16.30　The wash station must include hot/warm and cold running water with soap and towels. A receptacle must also be provided to enable the washing of face, hands and forearms.

16.31　A mess room will be required for periods of rest and the eating of meals. A source of potable water for drinking and a means of heating food and water must also be provided. This should be away from the main work area.

16.32　A drying room must be provided for drying workers' wet clothes and footwear where appropriate.

16.33　A suitable area should be made available for holding site meetings.

The CDM Co-ordinator will approve facilities *before* commencement of works on site.

Working at heights

16.34　The Principal Contractor shall ensure full compliance with the Work at Heights Regulations 2005 and shall ensure a comprehensive management system is in place for managing site access equipment for all contractors e.g. ladders, tower scaffolds and so on.

Access equipment

16.35　The Principal Contractor shall ensure that he has adequate management procedures in place for controlling the safe use of access equipment across the site.

16.36　Where the Principal Contractor provides access equipment for use by all other contractors he shall ensure that all checks and statutory inspections are carried out and that operators are competent to use the equipment.

16.37   Where contractors and sub-contractors provide their own access equipment for use on site it shall be checked by the Principal Contractor and, if necessary, its use restricted to named contractors only.

16.38   The Principal Contractor shall have in place procedures for preventing the use of unsafe access equipment and where necessary, shall remove it from site or disable it in a controlled way. Appropriate signage shall be displayed. Also, the Principal Contractor shall ensure that all users of *all* access equipment are competent to do so and where necessary, any sub-contractor shall be so advised that operators are not competent. Training shall be put in place.

17. Risk assessments and method statements

17.1   Risk Assessments as required by the Management of Health and Safety at Work Regulations 1999 and Method Statements are required to be undertaken and developed for all potentially hazardous works on site. The Principal Contractor shall collect and approve these documents from sub-contractors in addition to those created for his own works. Information and advice on the development and/or approval of Method Statements and Risk Assessments is obtainable from the CDM Co-ordinator.

17.2   The Project Architect shall create Design Risk Assessments prior to the construction phase of the project and these shall be made available, via the CDM Co-ordinator, to the Principal Contractor.

All other members of the Project Team (Structural Engineer, M&E Services Consultant, Principal Contractor, Sub-Contractor, etc.) shall develop Design Risk Assessments where they design or modify a design for any works, fixtures or fittings, etc. This is of particular note where any works are "design and build".

17.3   The Principal Contractor shall collect Designers Risk Assessments from all related parties, and forward these to the CDM Co-ordinator for inclusion within the Health and Safety File.

18. Commissioning and testing

18.1   In the later stages of the project, commissioning and testing of installed services and equipment is to be undertaken.

18.2   Items which will require commissioning include, but are not limited to:

- Lifts
- Fire alarms, fire detection equipment and fire suppression systems
- Emergency lighting
- Sprinkler systems
- Heating, ventilation and air conditioning equipment
- New incoming services
- Kitchen equipment
- Gas dispensing equipment
- Gas fires.

18.3   Commissioning works are to be undertaken by specialist contractors and all risks associated with these works are to be assessed and will be controlled by a permit to work system. The permit will refer to the specific procedures involved with "locking off", etc.

18.4   Specific risks associated with commissioning and testing work include:-

- 240 volt electricity
- Pressurized gas systems
- Hazardous substances (e.g. chlorine)
- Moving mechanical parts

18.5   Commissioning engineering are required to supply detailed risk assessments, method statements, permits and programme of works to the Principal Contractor prior to attendance on site to allow complete integration of health and safety on site.

19. Health and Safety File

19.1   The CDM Co-ordinator is responsible for the collation of the Health and Safety File.

19.2   The Principal Contractor shall collect all relevant information and forward it to the CDM Co-ordinator.

19.3   All information, surveys, reports and certificates are to be included with the formatted file. The CDM Co-ordinator will request relevant information from the Principal Contractor and/or from other Contractors as necessary.

19.4   The Project Team must provide all relevant documentation and information to the CDM Co-ordinator within 21 days of the contract completion.

19.5   The following information (where appropriate) is required:

- Timber and damp survey and certificate/guarantee, to include schedule of materials used.
- Details of any hazardous materials left on site.
- Clean air certificates following removal.
- Drainage test report.
- Structural survey report.
- Structural hazard assessment.
- Structural specification.
- Designers risk assessment.
- Planning approvals.
- "As built" drawings including structure, construction mechanical and electrical layout, drainage runs.
- Location of services and access hatches.
- Location of plant and equipment and motor rooms.
- Location of all emergency equipment, fire alarms, security alarms, isolation switches, sprinkler systems, gas shut off valves.
- Kitchen commissioning certificates.

- Lift/hoist test commissioning certificate.
- Electrical test certificates.
- Mechanical test certificates.
- Gas commissioning certificates.
- Fire alarm test certificate.
- Smoke/heat detectors commissioning certificates.
- Details of hazardous areas of the building i.e. fragile roofing, boiler rooms, confined spaces.
- Operating and maintenance manuals.
- Evidence that the manager has received training and instruction on all installed equipment.
- Building Control completion certificate.
- Pipework schematics.

19.6   The following equipment requires statutory examination/planning maintenance. The Architect is responsible for ensuring that the Client's Maintenance Department are aware that the equipment is within the premises and requires maintenance:

- Steam generating pressure systems (6 monthly)
- Air conditioning and refrigeration systems
- Kitchen extract systems
- Gas fires (annually)
- Boilers (annually)
- Electrical wiring systems (5 yearly)
- Eye bolts (6 monthly)
- Lifting equipment: passenger lifts (6 monthly)
                                 goods/food hoists (annually)
- Real fires: chimney sweeping (annually)

This list is not exhaustive and any other information of relevance should be provided, within 21 days of contract completion.

Appendix 11A

CONSTRUCTION PHASE HEALTH AND SAFETY PLAN

GENERAL INFORMATION

1.0 General description of project

2.0 Project details

Address of Project...

....

Client...

Architect. ....

Quantity Surveyor ....

Structural Engineer ....

M&E Consultant ....

CDM Co-ordinator ....

Principal Contractor ....

3.0 Project timescales

Lead-in time for design development ....

Lead-in time for site set up etc...

Start on site date...

Completion date...

Partial handover dates..

4.0 Notification and display of F10

When was F10 completed and sent to HSE...

Where will F10 be displayed on site...

5.0 Extent and location of existing information

List all information available to the Principal Contractor, location and ease of availability e.g. existing as built/installed drawings, existing O&M manuals.

HEALTH AND SAFETY MANAGEMENT

6.0 Health and safety objectives for the project

7.0 On-site organization and responsibility for health and safety

Site Agent...

Contracts/ProjectManager..

Health and Safety Manager..

Safety Director..

Principal Contractor's own Health and Safety Consultants (if any)..........

8.0 Who will have overall responsibility for on-site health and safety

9.0 Please detail their relevant qualifications and experience of managing health and safety within the construction industry

10.0 How will health and safety be managed on site

11.0 How will contractors and others be encouraged to discuss and communicate with each other on health and safety issues

12.0 How will contractors and sub-contractors be selected and how will their competency and resources be assessed

13.0 What action will be taken if contractors are deemed to be subsequently failing to do their jobs safely, ignoring site rules, etc.

14.0 How will a multi-occupied site be managed i.e. how will the principal contractor manage the inter-relationship between others at work within the premises who are not directly under his control. Please see the Pre-Construction Information Pack for information on client direct appointments

15.0 Who will undertake or approve risk/COSHH assessments for any activity/substance that may affect the health and safety of persons on site or within the vicinity of the site.

16.0 Who are designated "competent persons" for the site, including for tasks such as : demolition, electrical, cranes, etc.

17.0 Detail the procedures for assessing contractor's risk assessments and method statements

18.0 How will relevant health and safety information be communicated from the principal contractor to other contractors on site eg. notice-boards, tool box talks, meetings and so on

EMERGENCY PROCEDURES

19.0 First aid

How many trained first aiders are proposed for the site.

Names of trained first aiders and level of qualification e.g. First Aid Certificate (trained) or Appointed Person Certificate.

What requirements will contractors be required to fulfil in respect of first aid ie. provide own trained personnel.

Will the first aid facilities provided by the Principal Contractor be available to all contractors? How will PC ensure that these requirements are met?

Location and type of first aid kits.

Location of first aid room if any.

Name and address of emergency GP practice, A&E department, local hospital, etc.

N.B.: This information must be displayed prominently by all first aid kits and within the messing facilities.

20.0 Accident and incident reporting

How will accidents/incidents be recorded, reported and followed up.

Who will report accidents.

Who will investigate accidents/incidents.

What actions will be taken following an accident investigation and how will additional controls be implemented to prevent similar incidents re-occurring.

How will revised safe systems of work be communicated to all employees.

21.0 Fire safety

Has a separate Fire Safety Plan been attached to this Plan. YES/NO ......
....

If no, please describe the fire safety procedures you will adopt on the site, including details as follows:

Method of raising the alarm ....
....

Emergency lighting ....
....

Signage ....
....

Number and location of fire extinguishers ....
....

Means of escape ....
....

Assembly point ....
....

Smoking policy ....
....

Storage of flammable materials, substances ....
....

Use of LPG, hot works ....

....

Fire precautions in temporary accommodation ....

....

If yes, please ensure all the above are included in the Fire Safety Plan.

Number and names of Fire Wardens and details of relevant training and experience. If appropriate, please add to Fire Safety Plan.

What considerations have been given to co-ordinating construction site fire safety issues with those of adjoining premises, or with the Client's undertaking.

Who will carry out site specific Fire Risk Assessments, how will they be done and where will they be kept.

22.0 Other emergencies

Has consideration been given to any other emergencies which might occur during the course of the project. If so, to what extent and what procedures will be followed. Possible examples may include:

Structure collapse
Scaffold collapse
Chemical/substance release
Asbestos fibre release
Flood
Explosion
Gas leak

SITE WELFARE FACILITIES

23.0 Sanitary accommodation

Where will sanitary accommodation be located.

Complete the following table:

Number of operatives on site at any one time (max number to be used):

Of this number how many: male: female:

No. of WC's	No. of Urinals	No. of Wash Hand Basins

Will all facilities have:

Running water/flushing equipment	YES/NO
Hot and cold running water	YES/NO
Main drainage	YES/NO
Lighting	YES/NO
Ventilation	YES/NO
Hand drying facilities	YES/NO

Will any other facilities be provided e.g. showers, suitable wash stations for washing forearms, faces, etc.

24.0 What provision is to be made for eating meals, cooking and the provision of hot drinks

25.0 What drying room and outdoor clothing storage will be provided

26.0 How will a ready supply of drinking water be provided and where

27.0 What facilities will be provided for clients and site visitors and where will they be

28.0 Has consideration been given to the possibility of people with disabilities needing access to any facilities

ARRANGEMENTS FOR CONTROLLING SIGNIFICANT SITE RISKS

29.0 Site-specific risks

What are the significant health and safety risks associated with the project, the environment or the proposed construction works. (Refer to Pre-Construction Information Pack and Designers Risk Assessments.)

What control measures will be put in place to deal with them.

```



```

30.0 Services and utilities

Have all services, utilities been identified and information received regarding location, status, etc. e.g. overhead and underground cables, mains sewers, gas mains and so on. What are the locations of services and utilities.

```



```

If underground services have not already been identified, the Principal Contractor should consider how any ground penetrating operations (breaking through ground floor slab, excavations, other digging, etc.) will be managed and controlled, how services will be identified, trial digs, etc. Please attach a Method Statement.

```



```

What provision will be made for temporary services.

```



```

31.0 Vehicle access and transport routes

Describe vehicular access to the site, proposed routes for deliveries, etc.

```



```

How will operatives be kept separate from vehicles and if this is not possible, how will the risks be controlled. Identify if there's any need for excessive reversing/manoeuvring of vehicles.

Please attach Risk Assessments to the Plan.

32.0 Working at heights

What measures are proposed to reduce the risks for working at heights. Please attach Risk Assessments for your Plan.

If staircases are to be removed and a period of time exists before the new one arrives, please detail how access will be gained to upper floors and how materials will be taken to the upper floors.

33.0 Lifting operations

What measures are proposed to control lifting operations, when and how will cranes, hoists, MEWP's, etc. be used. Identify any significant risks to operation of cranes, etc. to persons/property in vicinity (e.g. overhead services, adjacent structures, narrow access, poor ground, and so on.)

34.0 Working on fragile materials

What measures are proposed to reduce the significant risks of falling through fragile materials. Please attach a Risk Assessment to your Plan.

35.0 COSHH

How will COSHH Assessments be checked and how will information be shared amongst all contractors and others working on the project.

What site-specific COSHH Assessments are expected.

How will hazardous substances be stored.

36.0 Working with electricity

What measures will be taken to ensure that mains voltage power is not a significant hazard within the construction site area.

What will be the maximum power rating for tools.

How will the commissioning and testing of equipment and plant be managed so as to ensure the safety of all site users.

```
┌─────────────────────────────────────────────┐
│                                             │
│                                             │
│                                             │
│                                             │
│                                             │
└─────────────────────────────────────────────┘
```

37.0 Dangerous/unsafe/unstable structures

What measures will be taken to prevent the collapse of structures other than in controlled sequences.

```
┌─────────────────────────────────────────────┐
│                                             │
│                                             │
│                                             │
│                                             │
│                                             │
└─────────────────────────────────────────────┘
```

OCCUPATIONAL HEALTH ISSUES

38.0 Asbestos

How will any discovery of asbestos or any other perceived or known harmful substance be managed on site.

```
┌─────────────────────────────────────────────┐
│                                             │
│                                             │
│                                             │
│                                             │
│                                             │
│                                             │
│                                             │
│                                             │
└─────────────────────────────────────────────┘
```

How will operatives be kept informed.

```
┌─────────────────────────────────────────────┐
│                                             │
│                                             │
└─────────────────────────────────────────────┘
```

What steps will be taken to prevent exposure to the risk of fibres, etc.

```
┌─────────────────────────────────────────────┐
│                                             │
│                                             │
└─────────────────────────────────────────────┘
```

39.0 Noise

How will exposure to excessive noise be controlled, especially to those working near noisy operations. How will the risks of excessive vibration be controlled.

40.0 Manual handling

How will all aspects of manual handling be controlled and the risk of back injury, repetitive strain injury, etc. be reduced on the site.

41.0 Contaminated land

How will exposure to any contaminated land be prevented, but if exposure does occur, how will it be managed.

42.0 Other health risks

Have any other health risks been identified within any project information, Design Risk Assessments, etc. e.g. exposure to cement dust and so on.

TRAINING

43.0 Site induction

How is it proposed to undertake and provide site induction training to operatives.

```



```

What subjects will be covered in induction training.

```



```

Please attach details of site induction programmes and subjects covered, etc. to the Plan.
How and what records will be kept of training.

```



```

What other training is envisaged.

```



```

How will Clients, design/project team members, visitors, etc. be inducted.

```



```

SITE SECURITY

44.0 Access/egress

What provisions are to be made for access and egress to the site, welfare facilities, etc. e.g. will site security and sign in desk be available, badges, ID cards and so on.

```



```

What will happen out of normal hours.

```
┌─────────────────────────────────────────────────────┐
│                                                     │
│                                                     │
│                                                     │
└─────────────────────────────────────────────────────┘
```

45.0 Prevention of unauthorized access

What measures are proposed to prevent the unauthorized access to the site, either during operating hours or outside them e.g. how will children be prevented from accessing site. Please refer to the Pre-Construction Information Pack for further guidance.

```
┌─────────────────────────────────────────────────────┐
│                                                     │
│                                                     │
│                                                     │
│                                                     │
└─────────────────────────────────────────────────────┘
```

ADJACENT LAND/BUILDINGS

46.0 What measures will be taken to protect the occupiers of adjacent premises from site activities.

```
┌─────────────────────────────────────────────────────┐
│                                                     │
│                                                     │
└─────────────────────────────────────────────────────┘
```

Do the neighbouring premises activities pose any risk to persons on site, if so, please detail specific measures required to protect all persons at risk.

```
┌─────────────────────────────────────────────────────┐
│                                                     │
│                                                     │
│                                                     │
└─────────────────────────────────────────────────────┘
```

DESIGN RISK ASSESSMENTS

47.0 Have design risk assessments been issued to the principal contractor by the designers? Have significant risks (if any) been discussed and planned into the work sequence? Give details

```
┌─────────────────────────────────────────────────────┐
│                                                     │
│                                                     │
│                                                     │
└─────────────────────────────────────────────────────┘
```

48.0 **How will information about design risk assessments be relayed to contractors on site**

49.0 **How will ongoing design changes be recorded, especially when designs are adapted or altered by contractors**

HEALTH AND SAFETY FILE

50.0 **How will information be collected, collated and passed to the CDM Co-ordinator during the progress of the project**

Any other information relevant:

Disclaimer

This proforma Construction Phase Health and Safety Plan format has been designed to assist the Principal Contractor in complying with Regulation 23 of the Construction (Design and Management) Regulations 2007. The layout and content of the Construction Phase Health and Safety Plan does not automatically infer acceptance by the Health and Safety Executive and requests for addition information and content may be made by any enforcing authority. The Principal Contractor is responsible for ensuring that he complies fully with the requirements of the CDM Regulations and other health and safety legislation, and in particular, the Principal Contractor must ensure that he has addressed site specific hazards and risks in a competent manner.

PERRY SCOTT NASH
associates limited

Perry Scott Nash House, 2 Arlington Court, Whittle Way, Arlington Business Park, Stevenage, SG1 2FS
t 01438 745 771   f 01438 745 772   e info@perryscottnash.co.uk   w www.perryscottnash.co.uk

Appendix 11B

CDM Regulations 2007
Construction Phase Plan Approval

Project:

Address:

Principal Contractor:

Date Plan received:

General overall comments on the content of the plan

Description of welfare facilities

Deemed adequate?　　　　　　　　　　　　　Yes ☐　　No ☐

If no, improvements required as follows:

Site management procedures

Are such arrangements deemed adequate?　　　Yes ☐　　No ☐

If no, improvements required as follows:

Competency and resources

Preparation and planning time allowed?

Competency procedures in place?　　　　　　Yes ☐　　No ☐

Is such a period deemed adequate?　　　　　Yes ☐　　No ☐

If no, increased time required is:

Authorisation to Commence Construction Works

Is the Construction Phase Health & Safety Plan deemed to meet the requirements of Regulations 22 and 23 CDM 2007?　　　　　Yes ☐　　No ☐

If no, what needs to be done to ensure compliance?

Are construction works authorised to start on site?　　Yes ☐　　No ☐

Has the Client authorised the CDM Co-ordinator to act on his/her behalf in respect of complying with Regulation 16 CDM 2007?　Yes ☐　　No ☐

Declaration of Approval

In our role as CDM Co-ordinator we, have advised the Client, as required under Regulation 20, that the Construction Phase Health & Safety Plan is suitable and sufficient and that as required by Regulation 16, they may authorise the commencement of construction works on site. We hereby authorise such works on their behalf.

Works can commence on site no sooner than:

Signed   Date:

Issued to: ☐ Client
                ☐ Principal Contractor
                ☐ Architect/Designers
                ☐ Contractors
                ☐ Others – please list:

Appendix 12A

PERRY SCOTT NASH
associates limited

Perry Scott Nash House, 2 Arlington Court, Whittle Way, Arlington Business Park, Stevenage, SG1 2FS
t 01438 745 771 f 01438 745 772 e info@perryscottnash.co.uk w www.perryscottnash.co.uk

CONTRACTOR COMPETENCY
QUESTIONNAIRE
Main/Principal Contractors/Contractors

Our Client requires that your competency in respect of health and safety should be assessed. The following information is required and should be completed as fully as possible.

Please add additional pages as and where necessary, stating the question number you are answering.

About Your Organisation

---

1. Name and Address of Organisation

   ....

   ....

   ....

   Post Code:..

   Telephone No.:...

2. Nature of Organisation

   ....

   ....

3. Name of Head/Senior Person responsible for overseeing health and safety within your Organisation (please print name).

   ....

   ....

What health and safety qualifications does this individual hold?

....

....

....

Is the named solely responsible for health and safety or do they hold other positions within the Organisation? Please state.

....

....

....

If the position is multifunctional how much time do they allocate to health and safety issues per month?

....

4. How do you employ your staff? For example, are staff employed on a sub-contracted basis, if so, in which positions? Are staff on temporary or permanent contracts?

....

....

....

Do you employ agency workers?

....

5. Membership of Professional Bodies:
    a) Corporate Membership

....

....

....

b) Individual(s) Membership

....

....

....

c) Corporate Accreditation

....

....

....

Are you quality assured following an accredited quality system? If so, please state which system and attach certificate of accreditation.

....

If you are not accredited, do you follow an internal quality management system? If so, please attach a copy of your Quality Manual contents page and quality statement.

....

Managing Safety

6. Do you have a health and safety policy?   YES/NO
    If yes, please attach a copy.

    Do you issue an employee handbook? YES/NO

    If yes, please attach a copy

    How do you emphasise management to health and safety within your organisation?
    ....
    ....
    ....
    ....

7. Have you ever been subject to any statutory action in relation to health and safety as an organisation or has any individual within the organisation? To include improvement notices, prohibition notices, prosecutions or cautions and also, civil claims. If the answer if yes, please provide details of the action taken, and measures taken to prevent a recurrence.
    ....
    ....
    ....
    ....
    ....

    Have you had any RIDDOR incidents within the last 5 years? If the answer is "yes", please provide details and measures taken.
    ....
    ....
    ....
    ....
    ....
    ....

8. Who are the "competent" person(s) within your Organisation with regard to the Management of Health and Safety Regulations 1999?
    ....
    ....
    ....

9. Do you rely on in-house or external expertise in regards to a health and safety competent person?
    ....

If external, please give name, address and telephone number of external organisation.

....

....

....

....

If internal, please give name and position within the organisation and provide a CV of the individual named.

....

....

....

How often do you have access to this advice?

....

10. How do you obtain information in respect of health and safety to ensure that your organisation is kept up to date in respect of new legislation?

....

....

....

Do you have access to any electronic systems or regular publications to aid your organisation in health and safety updates? If so, please state which.

....

....

....

....

11. Active monitoring – please identify which form of active monitoring you undertake as an organisation, either internally or by another on your behalf.

> Safety Audit*
> Safety Tours*
> Safety Inspections*
> Safety Sampling*

How regularly are these undertaken and by who.

....

....

Pro-active monitoring – please identify which form of monitoring has been performed by yourself or others on your behalf.

> Ill-health reporting*
> Dangerous occurrence investigations*
> Near miss reporting*
> Accident statistics analysis*

Please state who undertakes this monitoring.

....

....

12. How do you manage health and safety where you are Principal Contractor on a multi-occupied site? E.g. where the site is also occupied by a Client as in a building extension or as in a shopping complex, where you may be one of many Principal Contractors with shared responsibilities.

....

....

....

....

....

Risk Assessments and Method Statements

13. What procedures do you have for completing Risk Assessments?

....

....

....

What training do operatives have to ensure they understand the need of *site specific* risk assessments?

....

....

Who compiles Method Statements and how do you ensure that these are communicated to the work force?

....

....

....

What procedures do you adopt for training operatives on Method Statements and for ensuring that they understand their importance?

....

....

....

....

....

Managing Sub-Contractors

14. Do you sub-contract any activities? Please list.

....

....

15. How do you assess that a sub-contractor is competent in respect of health and safety?

    ....
    ....
    ....
    ....

    Do you use sub-contractor competency questionnaires?   YES/NO
    If so, please attach a copy

    Do you limit the number of sub-contractors which you will have on a site at any one time, if so, what limit do you set?

    ....
    ....
    ....

    How do you sign off sub-contracted work upon completion?

    ....
    ....
    ....

16. How do you ensure that sub-contractors work safely on site?

    ....
    ....

17. What action have you taken or would you take in respect of a sub-contractor following an unsafe system of work?

    ....
    ....
    ....

Maintenance of Equipment, Plant etc.

18. Is the equipment you use on site primarily your own or hired in?

    ....

    If hired in, which suppliers do you most regularly use and how do you assess their service?

    ....
    ....
    ....
    ....

    How do you assess the safety of electrical equipment used on site in respect of electrical safety and appropriate guarding?

    ....
    ....
    ....

Do you undertake routine maintenance of your site equipment, how often is this undertaken, by whom and what qualifications do they hold?

....

....

....

How are faults or damage to equipment reported and how are faulty items identified and removed from use?

....

....

....

Occupational Health

19. Do you have a formal procedure for completing COSHH assessments, if so please attach a copy.

....

Who undertakes COSHH assessments and how are these checked within your organisation? What qualification do they hold or what training have they received?

....

....

....

What systems do you have in place to try to eliminate the use of hazardous substances?

....

....

....

20. How do you eliminate or minimise manual handling activities on site?

....

....

....

What training do your operatives receive on manual handling? How do you convey information on manual handling to sub-contractors?

....

....

....

....

Who is responsible for completing manual handling risk assessments and how often are they done? (Please supply examples).

....

....

21. What approach do you have to noise control on site?

....

....

....

....

Who conducts noise assessments and what are their qualifications/ experience?

....

....

....

....

22. What approach do you have to advising your employees, operatives, sub-contractors on ill-health issues, such as dermatitis, sun burn, inhalation of dust?

....

....

....

....

23. What procedures do you adopt when asbestos is either known to be on site, or is found unexpectedly?

....

....

....

....

Who in your Company is trained in asbestos awareness and safe procedures?

....

....

Assessment and Management of Resources

24. Please provide details of resources available to you as a contractor. Include number of staff available for professional input, administration etc.

....

....

....

25. What is the maximum number of projects that you are able to work on at any one time? Please be as specific as possible.

....

....

26. What type of work do you consider to be your speciality?

   ....

   ....

27. Please would you describe what you believe to be some of the major concerns in respect of health and safety with a fast-track pub/restaurant development programme. What do you consider the key health and safety issues and how do you propose to manage them?

   ....

   ....

   ....

   ....

28. Are there any restrictions in respect of work you are able to undertake eg. distance from head office?

   ....

   ....

   ....

29. Please detail your approach to and management of snagging throughout and at the end of a project.

   ....

   ....

   ....

   ....

Training and Competency

---

30. How do you appoint site agents to a project?

   ....

   ....

   ....

31. What training do site agents and contracts managers receive in respect of health and safety?

   ....

   ....

   ....

32. How do you ensure that sub-contract employees and visitors on site are given specific information concerning that site?

   ....

   ....

   ....

33. What level of training is given to site operatives?

....

....

How many staff have received first aid training, which organisation did they receive their training through?

....

....

Construction Phase Health and Safety Plan

34. What procedures do you have for ensuring that this is site specific and relevant to the project in hand, eg. who completes it?

....

....

....

The following documentation must accompany the questionnaire:

35. Your Company Health and Safety Policy

36. Your Company Employees Handbook

37. Evidence of safety management procedures, including safety auditing procedures.

38. Provide three method statements from a contract you have undertaken in the last 12 months.

39. Provide three risk assessments from a contract you have undertaken in the last 12 months.

40. Provide three COSHH assessments from a contract you have undertaken in the last 12 months.

41. Provide three manual handling assessments from a contract you have undertaken in the last 12 months.

42. Provide *all* information on *any* criminal charges in relation to health and safety. This should include details of improvement or prohibition notices, prosecutions or cautions.

43. Provide information on any civil claims under Employer or Public Liability Insurance, or any other civil action claims.

44. Provide evidence of liability insurance.

45. Quality Assurance Manual (contents page) – if applicable.

Experience and References

46. Please describe your experiences in carrying out the following works:

"pub refurbishment works and pub/restaurant new development works"

Under CDM 2007 Regulations you must be able to demonstrate competency in the tasks you are appointed for. Please describe any similar works undertaken. (Please continue on a separate sheet if necessary).

....

....

....

47. Please provide name of two other Clients (including telephone numbers) outside of your organisation, who can act as referees and vouch for your competency in respect of health and safety.

Name of Company ....

Name of Contact and Position in Company ....

....

Telephone No. ....

Name of Company ....

Name of Contact and Position in Company ....

....

Telephone No. ....

Appendix 13A

HEALTH & SAFETY FILE

PART 1

CONSTRUCTION (DESIGN & MANAGEMENT) REGULATIONS 2007

Contents

Section E	Cleaning Work
Section F	Maintenance Work
Section G	Hazardous areas/equipment within the building
	Safety precautions to be followed
Section H	As Built Drawings (Construction)
	Schedule of Materials Used
	Schedule of Operating and Maintenance Manuals
	Schedule of Equipment with Health and Safety
	Implications
	As Installed Mechanical and Electrical Drawings
	Structural Specification and Drawings
	Miscellaneous Information

Please note

The information contained in the Health and Safety File which is kept on the premises must be read in conjunction with the Operating and Maintenance Manuals for Mechanical and Electrical also kept on the premises.

This Health and Safety File is a legal document and is required by law to be made available to any person who requires information relating to the construction of the building.

It must not be altered or amended without authorization.

Section A

Introduction

The Construction (Design and Management) Regulations 2007 require the provision of a Health and Safety File for all new buildings.

The Health and Safety File should contain:

(i)   Information included with the design by virtue of Regulation 11
(ii) Any other information relating to the project which it is reasonably foreseeable will be necessary to ensure the health and safety of any person at work who is carrying out or will carry out construction, maintenance and cleaning work in relation to the project or any person at work who may be affected by the work of such person.

Regulation 18

The premises at ************************, known as ****************, operates as a **************** and the general arrangements, plant and equipment within the premises are common to the design and operation of all ****************** premises. There are no design criteria, specification criteria, etc. which cause any concern in respect of health and safety issues.

Provided all usual health and safety procedures and existing health and safety legislation is complied with there are no known reasons why the health and safety of any *competent* person should be compromised when they are undertaking any maintenance, cleaning or future construction work within the building.

Section A

LOCATION OF PREMISES:

CLIENT:

Tel:

DESIGNERS
Architects:

Tel:

Quantity Surveyors:

Tel:

Structural Engineers:

Tel:

Building Services Consultants:

Tel:

CDM Co-ordinator:

Tel:

Principal Contractor:

Tel:

Mechanical Services Contractor:

Tel:

Electrical Services Contractor:

Tel:

Section A

Internal Fit out of developers newly constructed "shell building" including installation of mechanical and electrical services, to form a
****

[OR]

Complete refurbishment, including internal demolition, structural alterations and ancillary works of existing commercial units to form a
****

Construction timetables

Commencement on Site:
Completion:

Section B

Preliminary survey information

(a) Contaminated Land Search

A comprehensive site search was undertaken by ************* ********** and "No evidence was established that the Client's site has been put to uses which would be classified as potentially contaminative".
(Report of findings attached).

[OR]

No contaminated land search was undertaken prior to the development project.

No evidence was found during preliminary enquiries that the development site has been put to uses which would be classified as potentially contaminative.

(b) Asbestos Survey

A comprehensive Asbestos Survey was carried out by ************* ******** and asbestos insulation, asbestos cement and asbestos insulating board were identified throughout the building.

[Client's Name] commissioned a licensed asbestos contractor to remove all asbestos as identified in the report.

All easily accessible and visible asbestos was removed.

However, as with all surveys, a complete guarantee that the building is totally free of asbestos cannot be given as unaccessed areas may still contain asbestos.

Summary of asbestos found and removed is attached.

[OR]

No asbestos survey was undertaken during the preliminary enquiries regarding the site. All asbestos found during the course of the construction project was removed or encapsulated. Please check Section C of this file for details.

The Client and Design Team cannot give a complete guarantee that the premises is free of asbestos or that all asbestos present has been located and identified. All personnel are to proceed therefore with due caution.

Section C

Existing materials/hazardous substances

Section D

(a) **Construction Work**

(b) **Structural Information**

(c) **Mechanical and Electrical Services**

Section E

Cleaning work

Section F

Maintenance work

The following equipment will be subject to periodic maintenance:

(i)   Electrical installation
(ii)   Fire alarm system
(iii)   Ventilation/mechanical plant
(iv)   Water tanks
(v)   Gas installation, pipework and meters
(vi)   Gas flame effect fire
(vii)   Boilers

Procedures for maintenance are included in the O&M Manuals as supplied by the Mechanical and Electrical Contractors and/or Building Services Consultants.

As already stated, ductwork may be subject to periodic cleaning.

All work undertaken on plant or equipment shall be undertaken with cognisance of:

The Electricity at Work Regulations 1989
The Gas (Installation and Use) Regulations 1998
The Provision and Use of Work Equipment Regulations 1998
The Personal Protective Equipment Regulations 1992

And of course, general health and safety legislation.

Plant and equipment is located ....

Condensers are sited ....

Section G

Hazardous areas/equipment

Section H

H.1   As Built Drawings (Construction)
H.2   Schedule of Materials Used
H.3   Schedule of Operating and Maintenance Manuals
H.4   Schedule of Equipment with Health and Safety Implications
H.5   As Installed Mechanical and Electrical Drawings
H.6   Structural Specification and Drawings
H.7   Miscellaneous Information

Appendix 15A

Site Safety Checklist

1. Safe Places of Work

- Can everyone reach their place of work safely, e.g. are roads, gangways, passageways, passenger hoists, staircases, ladders and scaffolds in good condition?
- Are there guard rails or equivalent protection to stop falls from open edges on scaffolds, mobile elevating work platforms, buildings, gangways, excavations, etc.?
- Are holes and openings securely guard railed, provided with an equivalent standard of edge protection or provided with fixed, clearly marked covers to prevent falls?
- Are structures stable, adequately braced and not overloaded?
- Are all working areas and walkways level and free from obstructions such as stored material and waste?
- Is the site tidy, and are materials stored safely?
- Are there proper arrangements for collecting and disposing of waste materials?
- Is the work adequately lit? Is sufficient additional lighting provided when work is carried on after dark or inside buildings?

2. Emergency/Fire Procedures

General

- Have emergency procedures been developed, e.g. evacuating the site in case of fire or rescue from a confined space?
- Are people on site aware of the procedures?
- Is there a means of raising the alarm and does it work?
- Are there adequate escape routes and are these kept clear?

Fire

- Is the quantity of flammable material on site kept to a minimum?
- Are there proper storage areas for flammable liquids and gases, e.g. LPG and acetylene?
- Are containers and cylinders returned to these stores at the end of the shift?
- If liquids are transferred from their original containers are the new containers suitable for flammable materials?
- Is smoking banned in areas where gases or flammable liquids are stored and used? Are other ignition sources also prohibited?

- Are gas cylinders and associated equipment in good condition?
- When gas cylinders are not in use, are the valves fully closed?
- Are cylinders stored outside?
- Are adequate bins or skips provided for storing waste?
- Is flammable and combustible waste removed regularly?
- Are the right number and type of fire extinguishers available and accessible?

3. Welfare

- Have suitable and sufficient numbers of toilets been provided and are they kept clean?
- Are there clean wash basins, warm water, soap and towels?
- Is suitable clothing provided for those who have to work in wet, dirty or otherwise adverse conditions?
- Are there facilities for changing, drying and storing clothes?
- Is drinking water provided?
- Is there a site hut or other accommodation where workers can sit, make tea and prepare food?
- Is there adequate first aid provision?
- Are welfare facilities easily and safely accessible to all who need to use them?

4. Protective Clothing

- Has adequate personal protective equipment, e.g. hard hats, safety boots, gloves, goggles, and dust masks been provided?
- Is the equipment in good condition and worn by all who need it?

5. Training and Induction

- Suitable induction training and records maintained.
- Trained/competent personnel.

6. F10/Health and Safety Plan

- F10 displayed?
- Construction Phase Safety Plan on site?

7. Accidents

- Accident book kept on site?
- Any reportable accidents?

8. Risk Assessments

- Have assessments been carried out for hazardous activities?
- Where appropriate, are records being kept?

9. Method Statements

- Are method statements in place for hazardous or unusual activities?
- Are the method statements being followed?

10. Hazardous Substances

- Have all harmful materials, e.g. asbestos, lead, solvents, paints etc. been identified?
- Have the risks to everyone who might be exposed to these substances been assessed?
- Have precautions been identified and put in place, e.g. is protective equipment provided and used; are workers and others who are not protected kept away from exposure?

11. Plant and Machinery

- Is the right plant and machinery being used for the job?
- Are all dangerous parts guarded, e.g. exposed gears, chain drives, projecting engine shafts?
- Are guards secured and in good repair?
- Is the machinery maintained in good repair and are all safety devices operating correctly?
- Are all operators trained and competent?

12. Lighting

- Is there sufficient light in all working areas?
- Are hazardous areas/escape routes suitably lit?

13. Electricity

- Is the supply voltage for tools and equipment the lowest necessary for the job (could battery operated tools and reduced voltage systems, e.g. 110 Volt, or even lower in wet conditions, be used)?
- Where mains voltage has to be used, are trip devices, e.g. residual current devices (RCDs) provided for all equipment?
- Are RCDs protected from damage, dust and dampness and checked daily by users?

- Are cables and leads protected from damage by sheathing, protective enclosures or by positioning away from causes of damage?
- Are all connections to the system properly made and are suitable plugs used?
- Is there an appropriate system of user checks, formal visual examinations by site managers and combined inspection and test by competent persons for all tools and equipment?
- Are scaffolders, roofers, etc. or cranes or other plant, working near or under overhead lines? Has the electricity supply been turned off, or have other precautions, such as "goal posts" or taped markers been provided to prevent them contacting the lines?
- Have underground electricity cables been located (with a cable locator and cable plans), marked, and precautions for safe digging been taken?

14. Scaffolds

- Are scaffolds erected, altered and dismantled by competent persons?
- Is there safe access to the scaffold platform?
- Are all uprights provided with base plates (and, where necessary, timber sole plates) or prevented in some other way from slipping or sinking?
- Are all the uprights, ledgers, braces and struts in position?
- Is the scaffold secured to the building or structure in enough places to prevent collapse?
- Are there adequate guard rails and toe boards or an equivalent standard of protection at every edge from which a person could fall 2.0 metre or more?
- Where guard rails and toe boards or similar are used:
  - are the toe boards at least 150 millimetre in height?
  - is the upper guard rail positioned at a height of at least 910 millimetre above the work area?
  - are additional precautions, e.g. intermediate guard rails or brick guards in place to ensure that there is no unprotected gap of more than 470 millimetre between the toe board and upper guard rail?
- Are the working platforms fully boarded and are the boards arranged to avoid tipping or tripping?
- Are there effective barriers or warning notices in place to stop people using an incomplete scaffold, e.g. where working platforms are not fully boarded?
- Has the scaffold been designed and constructed to cope with the materials stored on it and are these distributed evenly?

- Does a competent person inspect the scaffold regularly, e.g. at least once a week; always after it has been substantially altered, damaged and following extreme weather?
- Are the results of inspections recorded?

15. Demolitions

- Risk assessments/method statements?
- Restricted access and safe distances?
- Noise/dust control?
- PPE?

16. Cranes and Lifting Appliances

- Is the crane on a firm level base?
- Are the safe working loads and corresponding radii known and considered before any lifting begins?
- If the crane has a capacity of more than 1 tonne, does it have an automatic safe load indicator that is maintained and inspected weekly?
- Are all operators trained and competent?
- Has the banksman/slinger been trained to give signals and to attach loads correctly?
- Do the operator and banksman find out the weight and centre of gravity of the load before trying to lift it?
- Are cranes inspected weekly, and thoroughly examined every 14 months by a competent person?
- Are cranes inspected weekly, and thoroughly examined every 14 months by a competent person?
- Are the results of inspections and examinations recorded?
- Does the crane have a current test certificate?

17. Excavations

- Is an adequate supply of timber, trench sheets, props or other supporting material made available before excavation work begins?
- Is this material strong enough to support the sides?
- Is a safe method used for putting in the support, i.e. one that does not rely on people working within an unsupported trench?
- If the sides of the excavation are sloped back or battered, is the angle of batter sufficient to prevent collapse?
- Is there safe access to the excavation, e.g. by a sufficiently long, secured ladder?
- Are there guard rails or other equivalent protection to stop people falling in?

- Are properly secured stop blocks provided to prevent tipping vehicles falling in?
- Does the excavation affect the stability of neighbouring structures?
- Are materials, spoil or plant stored away from the edge of the excavation in order to reduce the likelihood of a collapse of the side?
- Is the excavation inspected by a competent person at the start of every shift; and after any accidental collapse or event likely to have affected its stability?

18. Roof Work

- Are there enough barriers and is there other edge protection to stop people or materials falling from roofs?
- Do the roof battens provide safe hand and foot holds? If not, are crawling ladders or boards provided and used?
- During industrial roofing, are precautions taken to stop people falling from the leading edge of the roof or from fragile or partially fixed sheets which could give way?
- Are suitable barriers, guard rails or covers, etc. provided where people pass or work near fragile material such as asbestos cement sheets and rooflights?
- Are crawling boards provided where work on fragile materials cannot be avoided?
- Are people excluded from the area below the roof work? If this is not possible, have additional precautions been taken to stop debris falling onto them?

19. Powered Access Equipment

- Has the equipment been erected by a competent person?
- Is fixed equipment, e.g. mast climbers, rigidly connected to the structure against which it is operating?
- Does the working platform have adequate guard rails and toe boards or other barriers to prevent people and materials falling off?
- Have precautions been taken to prevent people being struck by the moving platform, projections from the building or falling materials, e.g. barrier or fence around the base?
- Are the operators trained and competent?
- Is the power supply isolated and the equipment secured at the end of the working day?

20. Ladders

- Are ladders the right means of access for the job?
- Are all ladders in good condition?

- Are they secured to prevent them slipping sideways or outwards?
- Do ladders rise a sufficient height above their landing place? If not, are there other hand-holds available?
- Are the ladders positioned so that users do not have to over-stretch or climb over obstacles to work?
- Does the ladder rest against a solid surface and not on fragile or insecure materials?

21. Manual Handling

- Has the risk of manual handling injuries been assessed?
- Are hoists, telehandlers, wheel-barrows and other plant or equipment used so that manual lifting and handling of heavy objects is kept to a minimum?
- Are materials such as cement ordered in 25 kilogram bags?
- Can the handling of heavy blocks be avoided?

22. Hoists

- Is the hoist protected by a substantial enclosure to prevent someone from being struck by any moving part of the hoist or falling down the hoistway?
- Are gates provided at all landings, including ground level?
- Are gates kept shut except when the platform is at the landing?
- Are the controls arranged so that the hoist can be operated from one position only?
- Is the hoist operator trained and competent?
- Is the hoist safe working load clearly marked?
- If the hoist is for materials only, is there a warning notice on the platform or cage to stop people riding on it?
- Is the hoist inspected weekly, and thoroughly examined every six months by a competent person?
- Are the results of inspection recorded?

23. Noise/Dust

- Are breakers and other plant or machinery fitted with silencers?
- Are barriers erected to reduce the spread of noise?
- Is work sequenced to minimise the number of people exposed to noise?
- Are others not involved in the work kept away?
- Is suitable hearing protection provided and worn in noisy areas?
- Is dust being contained/kept to a minimum?

24. Protection of the Public

- Are the public fenced off or otherwise protected from the work?
- When work has stopped for the day:
  - are the gates secured?
  - is the perimeter fencing secure and undamaged?
  - are all ladders removed or their rungs boarded so that they cannot be used?
  - are excavations and openings securely covered or fenced off?
  - is all plant immobilised to prevent unauthorised use?
  - are bricks and materials safely stacked?
  - are flammable or dangerous substances locked away in secure storage places?

25. Hazardous Areas

- Suitably signed?
- Restricted access?
- Protected/guarded?

26. Confined Spaces

- Have any confined spaces been identified?
- Is a permit to work system in operation?

27. Hot Works

- Fire precautions?
- Permit to work system?

28. Traffic and Vehicles

- Have separate pedestrian, vehicles access points and routes around the site been provided? If not, are vehicles and pedestrians kept separate wherever possible?
- Have one-way systems or turning points been provided to minimise the need for reversing?
- Where vehicles have to reverse, are they controlled by properly trained banksmen?
- Are vehicles maintained; do the steering, handbrake and footbrake work properly?
- Have drivers received proper training?
- Are vehicles securely loaded?
- Are passengers prevented from riding in dangerous positions?

Appendix 16A

ACCIDENT & INCIDENT FORM

Premises

Address

Your Name   Job Title

Date reported to you

Name & address of
Person involved

Was that person an   Employee / Visitor / Other:

Male / Female   Male / Female   Age

If an employee,
what is their job title?

Details of what the
person involved
said happened

Was the person
involved injured?
If so, how?

Date & time of
accident or incident

If employee,
is that person
expected to be
off work or on
light duties for
more than
3 days?

Where did the
accident/incident
take place?

Did the person
go to hospital?

Were the police
involved?

Signed: _____

Date: _____

Appendix 21A

Excavation Checklist		
Location of Excavation		
1.	Have all underground pipes, utilities, etc. been identified and located?	YES/NO/NA
2.	Has an adequate supply of timber, trench sheets, props and other supporting material been delivered to site prior to excavation work beginning?	YES/NO/NA
3.	Have Risk Assessments and Method Statements been provided?	YES/NO/NA
4.	Is the material and method of support chosen, suitable for supporting the sides?	YES/NO/NA
5.	Is the method chosen for putting in the timbering safe ie. negates the need for persons to work in an unsupported trench?	YES/NO/NA
6.	Is the angle of any slope or batten sufficient to prevent collapse or land slip?	YES/NO/NA
7.	Is there safe access and egress to the excavation e.g. fixed ladders, tiered steps?	YES/NO/NA
8.	Are there barriers all around the excavation to stop people falling in?	YES/NO/NA
9.	Does the excavation affect the stability of any neighbouring buildings?	YES/NO/NA
10.	Are materials stacked safely, away from any edges and not likely to cause collapse of the excavation sides or slip into the excavation?	YES/NO/NA
11.	Is any plant or materials stored near the excavation sides so as to destabilise the excavation walls, etc.?	YES/NO/NA
12.	If vehicles tip into the excavation, are proper vehicle stops used, banksmen or other safe method of working to prevent the vehicle tipping into the excavation?	YES/NO/NA

13.	Are there any risks of fumes or other noxious or hazardous fumes, vapours or mists drifting into the excavation causing operatives to be affected?	YES/NO/NA
14.	Is there an emergency plan for evacuation or dealing with unexpected events?	YES/NO/NA
15.	Are all operatives trained and competent to work in the excavation?	YES/NO/NA
16.	Other matters – please describe	

Actions Required

Please indicate the steps which need to be taken to ensure that the excavation can be used safely.

Signed _____

Position _____

Company _____

Date _____

Time of Inspection _____

Appendix 23A

PERMIT TO WORK

Site Address _____

Site Agent _____

Brief description of the work and location
Sequence of work and control measures
Supervision arrangements
Individual responsible for controls and monitoring performance
Plant and equipment to be used and operator training requirements
Occupational Health Assessments (risk, noise, COSHH etc.)

Measures to ensure safety of third parties

Environmental controls

First aid and PPE requirements

Emergency procedures

Permit issued by _____

Permit issued to _____

Permit valid until _____

APPENDICES 24A

Ladder and Stepladder Checklist

Date:

Ladder Checklist
Ladder Identification Number:

Checked Area	Condition Good	Defects	Comments	Signature
Missing steps of rungs				
Loose steps or rungs (considered loose if they can be moved at all by hand)				
Loose nails, screws, bolts or other parts				
Cracked, split or very worn or broken stiles, braces, steps or rungs				
Slivers on stiles, rungs or steps				

Damaged or worn non-slip bases						
Twisted or distorted stiles						
Identification disc missing or illegible						
Extension Ladder Checklist **Ladder Identification Number:**						
Loose, broken or missing extension locks						
Defective locks that do not seat properly when ladder is expanded						
Rusted or corroded metal parts						
Worn or badly deteriorated cords						
Damaged, missing or seized pulleys						

Step Ladder Checklist					
Ladder Identification Number:					
Side strain (wobbly)					
Loose or bent hinge spreaders					
Stop on hinge spreader broken					
Broken, split or work steps					
Loose hinges					

APPENDICES 26A

CONTROL OF SUBSTANCES HAZARDOUS TO HEALTH REGULATIONS 2002

COMPANY:
ADDRESS:
CONTACT:

PRODUCT:

JOB TASK:

APPLICATION:

EQUIPMENT:

SAFETY DATA SHEET ATTACHED: YES   NO

RISK IDENTIFICATION
Hazardous Component(s):
Hazardous Nature of Component(s):
Health Hazards (Known):
Persons Affected:
Duration of Exposure:
Level of Exposure:
RISK CATEGORY:

CONTROL MEASURES

For Users:

For Persons in Location:

TRAINING:

HEALTH SURVEILLANCE:

RE-ASSESSMENT:

DATE OF ASSESSMENT/REVISION:

COSHH ASSESSMENT CARRIED OUT BY:

COMPANY:

APPENDICES 30A

FIRE RISK ASSESSMENT

Name and address of premises: ....

....

....

Owner/Employer/Person in Control: ....

Contact Details: ....

Date of Risk Assessment: ....

Completed by: ....

Use of premises/area under review: ....

Identification of Fire Hazards	High	Medium	Low

Identification and location of those at risk
Evaluation of the risks
Significant findings
Actions taken to reduce/remove risks

Residual Risk Assessment

High ☐   Medium/Normal ☐   Low ☐

Review of Risk Assessment:

Under what circumstances:

How often:

FIRE RISK ASSESSMENT

Name of Premises: _____

What particular area are you reviewing for this Fire Risk Assessment?

What activity, practice, operation etc are you particularly reviewing for this Fire Risk Assessment?

What ignition sources have you identified?

What sources of fuel have you identified?

Are there any 'extra' sources of oxygen, or will mechanical ventilation increase oxygen levels?

Does anyone do anything that will increase the risk of a fire starting?

If a fire were to start, who would be at risk?

Would anyone be at any extra or special risk, or would any injuries/ consequences of the fire be increased?

What precautions are currently in place to reduce the likelihood of a fire occurring, or to deal with it/control it if a fire did start e.g. checks, alarms, emergency procedures, etc.?

What other precautions need to be taken, if any: Does anything need to be done to improve existing fire precautions?

How will the information in this Risk Assessment be communicated to staff? Will any staff training take place?

Is there anything else that you think needs to be recorded on this Risk Assessment?

After having identified the hazards and risks of a fire starting within the premises and after considering all the procedures you *currently* have in place, do you consider the risk to *life safety* of either staff or customers (including any contractors, visitors, etc.) to be:

Very Likely ▢   Possible ▢   Unlikely ▢

If risks to life safety are very likely or possible, steps MUST be taken to improve fire safety.

If you implement the other additional measures identified in this Fire Risk Assessment, will risk to *life safety* of either staff or customers (plus others) be:

Very Likely ▢   Possible ▢   Unlikely ▢

If risks to life safety are possible or very likely, then greater control measures MUST be implemented.

Risk Assessment completed by: _____

Date: _____

Fire Risk Assessment needs a review on: _____

Glossary

Term	Meaning
CDM	Construction (Design and Management) Regulations 2007.
CDM Co-ordinator	A person appointed to carry out the duties as listed in Regulation 14 of the CDM 2007 Regulations.
Cleaning Work	The cleaning of any window or any transparent or translucent wall, ceiling or roof in or on a structure.
Client	Organization or individual for whom construction work is carried out or who carries out a project himself.
Competency	Demonstration by an individual or organization that they have sufficient experience, knowledge and other skills to carry out their duties satisfactorily.
Construction Phase	The period of time starting when construction work in any project starts and ending when construction work in that project is completed.
Construction Phase Health and Safety Plan	A safety plan which has to be developed before the commencement of the construction phase by the Principal Contractor and which must set out the health and safety management arrangements for the project so that the safety of all persons involved is protected.
Construction Site	Any place where construction work is being carried out or to which the workers have access.
Construction Work	Generally, the carrying out of building, civil engineering or engineering construction work. Regulation 2 CDM contains the full definition.
Contractor	An organization or individual who carries on a trade or business or other undertaking in connection with which he/she undertakes, carries out construction work, including sub-contractors.
Demolition/ Dismantling	The deliberate pulling down, destruction or taking apart of a structure or a substantial part of the structure, including dismantling and re-erection for re-use.
Design	Includes drawings, design details, specifications, bills of quantities, specification of articles and substances, which relate to a structure and calculations prepared for the purpose of a design.

(Continued)

Term	Meaning
Designer	Any individual or trade or business which involves them in preparing designs for construction work – e.g. preparing drawings, design details, specifications, bills of quantities, materials specifications.
Developer	Someone who arranges for construction works to be carried out whilst acting on behalf of a domestic client.
Domestic Client	People who have work done which does not relate to their trade or profession e.g. people commissioning building work on their own home.
Duty Holder	Someone who has duties under CDM Regulations 2007.
Excavation	Includes any earthworks, trench, well, shaft, tunnel or underground working.
Fragile Material	A surface or an assembly which is liable to give way if a person or load crosses it, works on it or which will collapse if a load is dropped on to it.
Hazard	Anything with the potential to cause harm to an individual, group of persons or damage to property.
Health and Safety File	Information generated during the course of the construction project which future owners and occupiers of the building may need to know from a health and safety point of view.
Hierarchy of Risk Control	The principle of risk management: (i) Eliminate the hazard completely (ii) Reduce the hazard to an acceptable level e.g. substitution (iii) Control of the hazard/risk at source (iv) Protect the individual (v) Monitor and review.
Information	Any information which it is reasonable to assume that an individual should know in order to ensure that they discharge their duties safely and protect the safety of others.
Maintenance	Repair, upkeep, redecoration of buildings, structures, plant and equipment. Includes cleaning with water or abrasives or cleaning with corrosive or toxic chemicals.
Notifiable Project	A project for which the construction phase will last more than 30 days or involve more than 500-person days, or construction work.
Place of Work	Any place which is used by any person at work for the purposes of construction work or for the purposes of any activity arising out of or in connection with construction work.

(Continued)

Term	Meaning
Pre-Construction Information Pack	All information on the construction project relevant to significant health and safety risks. Any environmental issues e.g. contaminated land should be included.
Principal Contractor	The Main or Managing Contractor for a construction project appointed by the Client to assume the duties of Principal Contractor. The contractor responsible for the overall health and safety management of a site.
Principles of Prevention	The steps required to be taken in order to eliminate risk or prevent the most serious of consequences.
Project	A project which includes or is intended to include construction work and includes all planning, design, management or other work involved in a project until the end of the construction phase.
Resources	A general term including availability of the necessary plant, equipment, technical expertise, trained and competent people and time with which to carry out construction projects and comply with CDM.
Risk	The likelihood of a hazard being realized.
Structure	Any building, timber, masonry, metal or re-inforced concrete structure, railway line or siding, tramway line, dock, harbour, inland navigation, tunnel, shaft, bridge, viaduct, waterworks, reservoir, pipe or pipeline, cable, aqueduct, sewer, sewage works, gasholder, road, airfield, sea defence works, river works, drainage works, earthworks, lagoon, dam, wall, caisson, mast, tower, pylon, underground tank, earth retaining structure, structure designed to preserve or alter any natural feature, fixed plant and any structure which could be similar to any of the above.
	Any formwork, falsework, scaffold or other structure designed or used to provide support or means of access during construction works.
	Structure will also mean any part of a structure.
Training	Formal instruction in health and safety matters e.g. practical demonstrations on how to use plant and equipment, training in hazard awareness and so on.
Workplace	A place, whether or not within a building or forming part of a building, structure or vehicle, where a person is to work, is working, for the time being works or customarily works, for gain or reward, and in relation to an employee, includes a place, or part of a place, under the control of the employer, (not being domestic accommodation provided for the employee).

Index